Wastewater Treatment and Management

Wastewater Treatment and Management

Edited by
Joe Sawyer

Larsen & Keller
www.larsen-keller.com

Wastewater Treatment and Management
Edited by Joe Sawyer
ISBN: 978-1-63549-697-0 (Hardback)

Larsen & Keller
Published by Larsen and Keller Education,
5 Penn Plaza,
19th Floor,
New York, NY 10001, USA

Cataloging-in-Publication Data

Wastewater treatment and management / edited by Joe Sawyer.
p. cm.
Includes bibliographical references and index.
ISBN 978-1-63549-697-0
1. Sewage--Purification. 2. Factory and trade waste--Purification.
3. Water reuse--Management. I. Sawyer, Joe.
TD745 .W37 2018
628.3--dc23

Printed and bound in China.

For more information regarding Larsen and Keller Education and its products, please visit the publisher's website www.larsen-keller.com

Table of Contents

Preface

Wastewater management refers to the process of converting polluted, damaged, stale, unsuitable water into reusable water. The processes and techniques used in this field are filtration, polishing, sedimentation, biochemical oxidation, phase separation, redox, etc. The topics covered in this extensive text deal with the core aspects of wastewater management. It provides detailed analysis of the different concepts and techniques used under this field. The textbook, with its detailed analyses and data, will prove immensely beneficial to professionals and students involved in this area at various levels.

A foreword of all Chapters of the book is provided below:

Chapter 1 - Rapid urbanization has resulted in many water-related problems, crisis and concerns. There has been a manifold increase in wastewater generation. It is disposed either partially untreated or sometimes even completely untreated. The untreated water percolates in the ground and contaminates the ground water, or it pollutes water bodies when released through a sewer. This calls for wastewater management. This is an introductory chapter which will introduce briefly all the significant aspects of wastewater management; **Chapter 2 -** Water pollutants can be categorized into seven types. They are inorganic pollutants (salts and metals), organic pollutants, radioactive pollutants, nutrients and agriculture runoff, pathogens, suspended solids and sediments (organic and inorganic) and thermal pollution. Sources of pollution is divided into point source, nonpoint source and groundwater pollution. The chapter strategically encompasses and incorporates the major components and key concepts of wastewater management, providing a complete understanding; **Chapter 3 -** Sewage needs to be treated properly before it is disposed. A few types of sanitation systems are combined sewer, effluent sewer, sanitary sewer, simplified sewerage, storm drain, and vacuum sewer. The various systems of maintaining sanitation are best understood in confluence with the major topics listed in the following section; **Chapter 4 -** Sewers are normally steeply inclined and are large in size to help free flow of huge amounts of sewage. A sewage system requires a proper design implementation to maximize its load and efficiency. It should be two to three metres below ground level and direct the sewage towards a lower area, and towards the treatment plant. The topics discussed in the chapter are of great importance to broaden the existing knowledge on wastewater management; **Chapter 5 -** Wastewater treatment is a method to convert wastewater into a product that can be disposed into water bodies so that it causes minimal harm. Wastewater goes through processes such as water treatment, stormwater management and sewage treatment. This chapter discusses the methods of various kinds of treatment methods in a critical manner providing key analysis to the subject matter.

I would like to thank the entire editorial team who made sincere efforts for this book and my family who supported me in my efforts of working on this book. I take this opportunity to thank all those who have been a guiding force throughout my life.

Editor

1

Understanding Wastewater Management

Rapid urbanization has resulted in many water-related problems, crisis and concerns. There has been a manifold increase in wastewater generation. It is disposed either partially untreated or sometimes even completely untreated. The untreated water percolates in the ground and contaminates the ground water, or it pollutes water bodies when released through a sewer. This calls for wastewater management. This is an introductory chapter which will introduce briefly all the significant aspects of wastewater management.

Wastewater Management

Urbanization has encouraged the migration of people from villages to the urban areas. This has given rise to a number of environmental problems such as, water supply with desirable quality and quantity, wastewater generation and its collection, treatment and disposal. In urban areas for domestic and industrial uses the source of water is generally reservoir, river, lake, and wells. Out of this total water supplied, generally 60 to 80% contributes as a wastewater. In most of the cities, wastewater is let out partially treated or untreated and it either percolates into the ground and in turn contaminates the ground water or it is discharged into the natural drainage system causing pollution in downstream water bodies.

The importance of water quality as a factor constraining water use has often gone unacknowledged in the analyses of water scarcity. Water scarcity is a function not only of volumetric supply, but also of quality sufficient to meet the demand. The drinking water demand is perhaps the largest demand for high quality water apart from many industrial uses which also require high quality water. Agriculture, by far the largest consumer of water, also suffers when water supplies become saline. In India, water pollution comes from the main sources such as domestic sewage, industrial effluents, leachets from landfills, and run-off from solid waste dumps and agriculture land. Domestic sewage (black water) and sullage (grey water) is the main source of water pollution in India, especially in and around large urban centers. The regular monitoring of the water quality in the rivers and wells in the country revealed that the total coliform counts far exceeds the desired level in water to be fit for human consumption.

In the past disposal of waste from water closets was carried out manually and wastewater generated from kitchen and bathrooms was allowed to flow along the open drains. This primitive method was modified and replace by a water carriage system, in which these wastes are mixed with sufficient quantity of water. This waste is carried through closed conduits under the conditions of gravity flow. This mixture of water and waste products is known as sewage.

The advantages offered by the water carriage system are:

- The carriage of wastes on head or carts is not required.
- Bad smell, which was unavoidable during open transport of sewage, is not occurring due to transport of this polluted water in closed conduits.
- The old system was posing the health hazards to sweepers and to the nearby residents, because of the possibilities of flies and insects transmitting disease germs from the accessible carts to the residents food eatables. This is avoided in water carriage system because of transport of night soil in close conduits.
- The human excreta is washed away as soon as it is produced in water carriage system, thus storing is not required as required in the old system of manual disposal. Thus, no bad smells are produced in closed conduit transport.
- In the old system, the wastewater generated from the kitchen and bathrooms was required to be carried through open roadside drains for disposal. This is avoided in sewerage system as the open drains could generate bad odours when used for disposal of organic wastes.
- The water carriage system does not occupy floor area, as the sewers are laid underground.
- In addition, the construction of toilets one above the other is possible in water carriage system and combining latrine and bathrooms together as water closets is possible. This is one of the important advantages of water carriage system.

However, this water carriage system also has certain drawbacks such as:

- A large network of pipes is required for collection of the sewage; hence, the capital cost for water carriage system is very high.
- In addition, the operation and maintenance of sewerage system is very expensive.
- Large wastewater volume is required to be treated before disposal.

Definitions

Industrial wastewater: It is the wastewater generated from the industrial and commercial areas. This wastewater contains objectionable organic and inorganic compounds that may not be amenable to conventional treatment processes.

Night Soil: It is a term used to indicate the human and animal excreta.

Sanitary sewage: Sewage originated from the residential buildings comes under this category. This is very foul in nature. It is the wastewater generated from the lavatory basins, urinals and water closets of residential buildings, office building, theatre and other institutions. It is also referred as domestic wastewater.

Sewage: It indicates the liquid waste originating from the domestic uses of water. It includes sullage, discharge from toilets, urinals, wastewater generated from commercial establishments, institutions, industrial establishments and also the groundwater and stormwater that may enter into the sewers. Its decomposition produces large quantities of malodorous gases, and it contains numerous pathogenic or disease producing bacteria, along with high concentration of organic matter and suspended solids.

Sewage Treatment Plant is a facility designed to receive the waste from domestic, commercial and industrial sources and to remove materials that damage water quality and compromise public health and safety when discharged into water receiving systems or land. It is combination of unit operations and unit processes developed to treat the sewage to desirable standards to suit effluent norms defined by regulating authority.

Sewer: It is an underground conduit or drain through which sewage is carried to a point of discharge or disposal. There are three types of sewer systems that are commonly used for sewage collection. Separate sewers are those which carry the house hold and industrial wastes only. Storm water drains are those which carry rain water from the roofs and street surfaces. Combine sewers are those which carry both sewage and storm water together in the same conduit. House sewer (or drain) is used to discharge the sewage from a building to a street sewer. Lateral sewer is a sewer which collects sewage directly from the household buildings. Branch sewer or submain sewer is a sewer which receives sewage from a relatively small area. Main sewer or trunk sewer is a sewer that receives sewage from many tributary branches and sewers, serving as an outlet for a large territory. Depressed sewer is a section of sewer constructed lower than adjacent sections to pass beneath an obstacle or obstruction. It runs full under the force of gravity and at greater than atmospheric pressure. The sewage enters and leaves the depressed sewer at atmospheric pressure. Intercepting sewer is a sewer laid transversely to main sewer system to intercept the dry weather flow of sewage and additional surface and storm water as may be desirable. An intercepting sewer is usually a large sewer, flowing parallel to a natural drainage channel, into which a number of main or outfall sewers discharge. Outfall sewer receives entire sewage from the collection system and finally it is discharged to a common point. Relief sewer or overflow sewer is used to carry the flow in excess of the capacity of an existing sewer.

Sewerage: The term sewerage refers the infrastructure which includes device, equipment and appurtenances for the collection, transportation and pumping of sewage, but excluding works for the treatment of sewage. Basically it is a water carriage system designed and constructed for collecting and carrying of sewage through sewers.

Stormwater: It indicates the rain water of the locality.

Subsoil water: Groundwater that enters into the sewers through leakages is called subsoil water.

Sullage: This refers to the wastewater generated from bathrooms, kitchens, washing place and wash basins, etc. Composition of this waste does not involve higher concentration of organic matter and it is less polluted water as compared to sewage.

Wastewater: The term wastewater includes both organic and inorganic constituents, in soluble or suspended form, and mineral content of liquid waste carried through liquid media. Generally the organic portion of the wastewater undergoes biological decompositions and the mineral matter may combine with water to form dissolved solids.

Sewage

Sewage is a water-carried waste, in solution or suspension, that is intended to be removed from a community. Also known as domestic or municipal wastewater, it is characterized by volume or rate of flow, physical condition, chemical and toxic constituents, and its bacteriologic status (which organisms it contains and in what quantities). It consists mostly of greywater (from sinks, tubs, showers, dishwashers, and clothes washers), blackwater (the water used to flush toilets, combined with the human waste that it flushes away); soaps and detergents; and toilet paper (less so in regions where bidets are widely used instead of paper). Whether it also contains surface runoff depends on the design of sewer system.

Raw sewage arriving at a sewage treatment plant in Syria

All sewage ends up back in the environment (from which its constituents came), by any of several routes. A basic distinction in its route is whether it undergoes sewage treatment to mitigate its effect on the environment before arriving there. Sewage usually travels from a building's plumbing either into a sewer, which will carry it elsewhere, or into an onsite sewage facility (of which there are many kinds). Whether it is combined with surface runoff in the sewer depends on the sewer design (sanitary sewer or combined sewer). Before the 20th century, sewers usually discharged into a body of water such as a stream, river, lake, bay, or ocean. There was no treatment, so the breakdown of the human waste was left to the ecosystem. Today, the goal is that sewers route their contents to a wastewater treatment plant rather than directly to a body of water. In many countries, this is the norm; in many developing countries, it may be a yet-unrealized goal. In general, with passing decades and centuries, humanity seeks to be smarter about the route of sewage on its way back to the environment, in order to reduce environmental degradation and achieve sustainability. Thus other goals of modern sewage routing include handling surface runoff separately from sewage, handling greywater separately from toilet waste, and coping better with abnormal events (such as peaks in use from internal displacement and peaks in stormwater volumes from extreme weather).

Proper collection and safe, nuisance-free disposal of the liquid wastes of a community are legally recognized as a necessity in an urbanized, industrialized society. The reality is, however, that most wastewater produced globally remains untreated causing widespread water pollution, especially in low-income countries: A global estimate by UNDP and UN-Habitat is that 90% of all wastewater generated is released into the environment untreated. In many developing countries the bulk of domestic and industrial wastewater is discharged without any treatment or after primary treatment only.

The term sewage is nowadays regarded as an older term and is being more and more replaced by "wastewater". In general American English usage, the terms "sewage" and "sewerage" mean the same thing. Both words are descended from Old French *assewer*, derived from the Latin *ex-aquare*, "to drain out (water)". In American technical and professional English usage, "sewerage" refers to the infrastructure that conveys sewage.

Types

Greywater samples before and after treatment in a constructed wetland in Peru

- The wastewater from residences and institutions, carrying bodily wastes (primarily feces and urine), washing water, food preparation wastes, laundry wastes, and other waste products of normal living, are classed as domestic or sanitary sewage.
- Liquid-carried wastes from stores and service establishments serving the immediate community, termed commercial wastes, are included in the sanitary or domestic sewage category if their characteristics are similar to household flows. Wastes that result from an industrial processes such as the production or manufacture of goods are classed as industrial wastewater, not as sewage.
- Surface runoff, also known as storm flow or overland flow, is that portion of precipitation that runs rapidly over the ground surface to a defined channel. Precipitation absorbs gases and particulates from the atmosphere, dissolves and leaches materials from vegetation and soil, suspends matter from the land, washes spills and debris from urban streets and highways, and carries all these pollutants as wastes in its flow to a collection point.

Pollutant

Organic Pollutants and Nutrients

Sewage is a complex mixture of chemicals, with many distinctive chemical characteristics. These include high concentrations of ammonium, nitrate, phosphorus, high conductivity (due to high dissolved solids), high alkalinity, with pH typically ranging between 7 and 8. The organic matter of sewage is measured by determining its biological oxygen demand (BOD) or the chemical oxygen demand (COD). Sewage also contains the nutrients nitrogen and phosphorus.

Pathogens and Micro-pollutants

Sewage can be monitored for both disease-causing and benign organisms with a variety of techniques. Traditional techniques involve filtering, staining, and examining samples under a microscope. Much more sensitive and specific testing can be accomplished with DNA sequencing, such as when looking for rare organisms, attempting eradication, testing specifically for drug-resistant strains, or discovering new species. Sequencing DNA from an environmental sample is known as metagenomics.

Sewage has also been analyzed to determine relative rates of use of prescription and illegal drugs among municipal populations.

Sewage also contains environmental persistent pharmaceutical pollutants.

Trihalomethanes can also be present as a result of past disinfection.

Health and Environmental Aspects

All categories of sewage are likely to carry pathogenic organisms that can transmit disease to humans and animals. Sewage also contains organic matter that can cause odor and attract flies.

Sewage contains nutrients that may cause eutrophication of receiving water bodies; and can lead to ecotoxicity.

Reuse of Sewage

Increasingly, agriculture is using untreated wastewater for irrigation. Cities provide lucrative markets for fresh produce, so are attractive to farmers. However, because agriculture has to compete for increasingly scarce water resources with industry and municipal users, there is often no alternative for farmers but to use water polluted with urban waste, including sewage, directly to water their crops. There can be significant health hazards related to using water loaded with pathogens in this way, especially if people eat raw vegetables that have been irrigated with the polluted water.

The International Water Management Institute has worked in India, Pakistan, Vietnam, Ghana, Ethiopia, Mexico and other countries on various projects aimed at assessing and reducing risks of wastewater irrigation. They advocate a 'multiple-barrier' approach to wastewater use, where farmers are encouraged to adopt various risk-reducing behaviours. These include ceasing irrigation a few days before harvesting to allow pathogens to die off in the sunlight, applying water carefully so it does not contaminate leaves likely to be eaten raw, cleaning vegetables with disinfectant or allowing fecal sludge used in farming to dry before being used as a human manure. The World Health Organization has developed guidelines for safe water use.

Collection and Disposal

A medieval waste pipe in Stockholm Old Town formerly deposited sewage on the street to be flushed away by rain.

Sewage canal of a medieval house as depicted in 1447 *St. Barbara Altarpiece* in the National Museum in Warsaw.

A system of sewer pipes (sewers) collects sewage and takes it for treatment or disposal. The system of sewers is called *sewerage* or *sewerage system* in British English and *sewage system* in American English. Where a main sewerage system has not been provided, sewage may be collected from homes by pipes into septic tanks or cesspits, where it may be treated or collected in vehicles and taken for treatment or disposal. Properly functioning septic tanks require emptying every 2–5 years depending on the load of the system.

Sewage and wastewater is also disposed of to rivers, streams, and the sea in many parts of the world. Doing so can lead to serious pollution of the receiving water. This is common in developing countries and may still occur in some developed countries, where septic tank systems are too expensive.

Treatment

Sewage treatment is the process of removing the contaminants from sewage to produce liquid and solid (sludge) suitable for discharge to the environment or for reuse. It is a form of waste management. A septic tank or other on-site wastewater treatment system such as biofilters or constructed wetlands can be used to treat sewage close to where it is created.

Sewage treatment results in sewage sludge which requires sewage sludge treatment before safe disposal or reuse. Under certain circumstances, the treated sewage sludge might be termed "biosolids" and can be used as a fertilizer.

In developed countries sewage collection and treatment is typically subject to local and national regulations and standards.

Water Pollution

Raw sewage and industrial waste in the New River as it passes from Mexicali to Calexico, California

Water pollution is the contamination of water bodies (e.g. lakes, rivers, oceans, aquifers and groundwater). This form of environmental degradation occurs when pollutants are directly or indirectly discharged into water bodies without adequate treatment to remove harmful compounds.

Water pollution affects the entire biosphere – plants and organisms living in these bodies of water. In almost all cases the effect is damaging not only to individual species and population, but also to the natural biological communities.

Introduction

Water pollution is a major global problem which requires ongoing evaluation and revision of water resource policy at all levels (international down to individual aquifers and wells). It has been suggested that water pollution is the leading worldwide cause of deaths and diseases, and that it accounts for the deaths of more than 14,000 people daily. An estimated 580 people in India die of water pollution related illness every day. About 90 percent of the water in the cities of China is polluted. As of 2007, half a billion Chinese had no access to safe drinking water. In addition to the acute problems of water pollution in developing countries, developed countries also continue to struggle with pollution problems. For example, in the most recent national report on water quality in the United States, 44 percent of assessed stream miles, 64 percent of assessed lake acres, and 30 percent of assessed bays and estuarine square miles were classified as polluted. The head of China's national development agency said in 2007 that one quarter the length of China's seven main rivers were so poisoned the water harmed the skin.

Pollution in the Lachine Canal, Canada

Water is typically referred to as polluted when it is impaired by anthropogenic contaminants and either does not support a human use, such as drinking water, or undergoes a marked shift in its ability to support its constituent biotic communities, such as fish. Natural phenomena such as volcanoes, algae blooms, storms, and earthquakes also cause major changes in water quality and the ecological status of water.

Categories

Although interrelated, surface water and groundwater have often been studied and managed as separate resources. Surface water seeps through the soil and becomes groundwater. Conversely, groundwater can also feed surface water sources. Sources of surface water pollution are generally grouped into two categories based on their origin.

Point Sources

Point source pollution – Shipyard – Rio de Janeiro.

Point source water pollution refers to contaminants that enter a waterway from a single, identifiable source, such as a pipe or ditch. Examples of sources in this category include discharges from a sewage treatment plant, a factory, or a city storm drain. The U.S. Clean Water Act (CWA) defines point source for regulatory enforcement purposes. The CWA definition of point source was amended in 1987 to include municipal storm sewer systems, as well as industrial storm water, such as from construction sites.

Non-point Sources

Nonpoint source pollution refers to diffuse contamination that does not originate from a single discrete source. NPS pollution is often the cumulative effect of small amounts of contaminants gathered from a large area. A common example is the leaching out of nitrogen compounds from fertilized agricultural lands. Nutrient runoff in storm water from "sheet flow" over an agricultural field or a forest are also cited as examples of NPS pollution.

Blue drain and yellow fish symbol used by the UK Environment Agency to raise awareness of the ecological impacts of contaminating surface drainage

Contaminated storm water washed off of parking lots, roads and highways, called urban runoff, is sometimes included under the category of NPS pollution. However, because this runoff is typically channeled into storm drain systems and discharged through pipes to local surface waters, it becomes a point source.

Groundwater Pollution

Interactions between groundwater and surface water are complex. Consequently, groundwater pollution, also referred to as groundwater contamination, is not as easily classified as surface water pollution. By its very nature, groundwater aquifers are susceptible to contamination from sources that may not directly affect surface water bodies, and the distinction of point vs. non-point source may be irrelevant. A spill or ongoing release of chemical or radionuclide contaminants into soil (located away from a surface water body) may not create point or non-point source pollution but can contaminate the aquifer below, creating a toxic plume. The movement of the plume, called a plume front, may be analyzed through a hydrological transport model or groundwater model. Analysis of groundwater contamination may focus on soil characteristics and site geology, hydrogeology, hydrology, and the nature of the contaminants.

Causes

The specific contaminants leading to pollution in water include a wide spectrum of chemicals, pathogens, and physical changes such as elevated temperature and discoloration. While many of the chemicals and substances that are regulated may be naturally occurring (calcium, sodium, iron, manganese, etc.) the concentration is often the key in determining what is a natural component of water and what is a contaminant. High concentrations of naturally occurring substances can have negative impacts on aquatic flora and fauna.

Oxygen-depleting substances may be natural materials such as plant matter (e.g. leaves and grass) as well as man-made chemicals. Other natural and anthropogenic substances may cause turbidity (cloudiness) which blocks light and disrupts plant growth, and clogs the gills of some fish species.

Many of the chemical substances are toxic. Pathogens can produce waterborne diseases in either human or animal hosts. Alteration of water's physical chemistry includes acidity (change in pH), electrical conductivity, temperature, and eutrophication. Eutrophication is an increase in the concentration of chemical nutrients in an ecosystem to an extent that increases the primary productivity of the ecosystem. Depending on the degree of eutrophication, subsequent negative environmental effects such as anoxia (oxygen depletion) and severe reductions in water quality may occur, affecting fish and other animal populations.

Pathogens

A manhole cover unable to contain a sanitary sewer overflow.

Poster to teach people in South Asia about human activities leading to the pollution of water sources

Fecal sludge collected from pit latrines is dumped into a river at the Korogocho slum in Nairobi, Kenya.

Disease-causing microorganisms are referred to as pathogens. Although the vast majority of bacteria are either harmless or beneficial, a few pathogenic bacteria can cause disease. Coliform bacteria, which are not an actual cause of disease, are commonly used as a bacterial indicator of water pollution. Other microorganisms sometimes found in surface waters that have caused human health problems include:

- *Burkholderiapseudomallei*
- *Cryptosporidium parvum*
- *Giardia lamblia*
- *Salmonella*
- *Norovirus* and other viruses
- *Parasitic worms including the Schistosoma type*

High levels of pathogens may result from on-site sanitation systems (septic tanks, pit latrines) or inadequately treated sewage discharges. This can be caused by a sewage plant designed with less than secondary treatment (more typical in less-developed countries). In developed countries, older cities with aging infrastructure may have leaky sewage collection systems (pipes, pumps, valves), which can cause sanitary sewer overflows. Some cities also have combined sewers, which may discharge untreated sewage during rain storms.

Pathogen discharges may also be caused by poorly managed livestock operations.

Organic, Inorganic and Macroscopic Contaminants

Contaminants may include organic and inorganic substances.

A garbage collection boom in an urban-area stream in Auckland, New Zealand.

Muddy river polluted by sediment.

Organic water pollutants include:

- Detergents
- Disinfection by-products found in chemically disinfecteddrinking water, such as chloroform
- Food processing waste, which can include oxygen-demanding substances, fats and grease
- Insecticides and herbicides, a huge range of organohalides and other chemical compounds
- Petroleum hydrocarbons, including fuels (gasoline, diesel fuel, jet fuels, and fuel oil) and lubricants (motor oil), and fuel combustion byproducts, from storm waterrunoff
- Volatile organic compounds, such as industrial solvents, from improper storage.
- Chlorinated solvents, which are dense non-aqueous phase liquids, may fall to the bottom of reservoirs, since they don't mix well with water and are denser.

    - Polychlorinated biphenyl (PCBs)
    - Trichloroethylene
- Perchlorate
- Various chemical compounds found in personal hygiene and cosmetic products
- Drug pollution involving pharmaceutical drugs and their metabolites

Inorganic water pollutants include:

- Acidity caused by industrial discharges (especially sulfur dioxide from power plants)
- Ammonia from food processing waste
- Chemical waste as industrial by-products
- Fertilizers containing nutrients--nitrates and phosphates—which are found in storm water runoff from agriculture, as well as commercial and residential use
- Heavy metals from motor vehicles (via urban storm water runoff) and acid mine drainage
- Secretion of creosote preservative into the aquatic ecosystem
- Silt (sediment) in runoff from construction sites, logging, slash and burn practices or land clearing sites.

Macroscopic pollution – large visible items polluting the water – may be termed "floatables" in an urban storm water context, or marine debris when found on the open seas, and can include such items as:

- Trash or garbage (e.g. paper, plastic, or food waste) discarded by people on the ground, along with accidental or intentional dumping of rubbish, that are washed by rainfall into storm drains and eventually discharged into surface waters
- Nurdles, small ubiquitous waterborne plastic pellets
- Shipwrecks, large derelict ships.

The Brayton Point Power Station in Massachusetts discharges heated water to Mount Hope Bay.

Thermal Pollution

Thermal pollution is the rise or fall in the temperature of a natural body of water caused by human influence. Thermal pollution, unlike chemical pollution, results in a change in the physical properties of water. A common cause of thermal pollution is the use of water as a coolant by power plants and industrial manufacturers. Elevated water temperatures decrease oxygen levels, which can kill fish and alter food chain composition, reduce species biodiversity, and foster invasion by new thermophilic species. Urban runoff may also elevate temperature in surface waters.

Thermal pollution can also be caused by the release of very cold water from the base of reservoirs into warmer rivers.

Transport and Chemical Reactions of Water Pollutants

Most water pollutants are eventually carried by rivers into the oceans. In some areas of the world the influence can be traced one hundred miles from the mouth by studies using hydrology transport models. Advanced computer models such as SWMM or the DSSAM Model have been used in many locations worldwide to examine the fate of pollutants in aquatic systems. Indicator filter-feeding species such as copepods have also been used to study pollutant fates in the New York Bight, for example. The highest toxin loads are not directly at the mouth of the Hudson River, but 100 km (62 mi) south, since several days are required for incorporation into planktonic tissue. The Hudson discharge flows south along the coast due to the coriolis force. Further south are areas of oxygen depletion caused by chemicals using up oxygen and by algae blooms, caused by excess nutrients from algal cell death and decomposition. Fish and shellfish kills have been reported, because toxins climb the food chain after small fish consume copepods, then large fish eat smaller fish, etc. Each successive step up the food chain causes a cumulative concentration of pollutants such as heavy metals (e.g. mercury) and persistent organic pollutants such as DDT. This is known as bio-magnification, which is occasionally used interchangeably with bio-accumulation.

A polluted river draining an abandoned copper mine on Anglesey

Large gyres (vortexes) in the oceans trap floating plastic debris. The North Pacific Gyre, for example, has collected the so-called "Great Pacific Garbage Patch", which is now estimated to be one hundred times the size of Texas. Plastic debris can absorb toxic chemicals from ocean pollution, potentially poisoning any creature that eats it. Many of these long-lasting pieces wind up in the stomachs of marine birds and animals. This results in obstruction of digestive pathways, which leads to reduced appetite or even starvation.

Many chemicals undergo reactive decay or chemical change, especially over long periods of time in groundwater reservoirs. A noteworthy class of such chemicals is the chlorinated hydrocarbons such as trichloroethylene (used in industrial metal degreasing and electronics manufacturing) and tetrachloroethylene used in the dry cleaning industry. Both of these chemicals, which are carcinogens themselves, undergo partial decomposition reactions, leading to new hazardous chemicals (including dichloroethylene and vinyl chloride).

Groundwaterpollution is much more difficult to abate than surface pollution because groundwater can move great distances through unseen aquifers. Non-porous aquifers such as clays partially purify water of bacteria by simple filtration (adsorption and absorption), dilution, and, in some cases, chemical reactions and biological activity; however, in some cases, the pollutants merely transform to soil contaminants. Groundwater that moves through open fractures and caverns is not filtered and can be transported as easily as surface water. In fact, this can be aggravated by the human tendency to use natural sinkholes as dumps in areas of karst topography.

There are a variety of secondary effects stemming not from the original pollutant, but a derivative condition. An example is silt-bearing surface runoff, which can inhibit the penetration of sunlight through the water column, hampering photosynthesis in aquatic plants.

Measurement

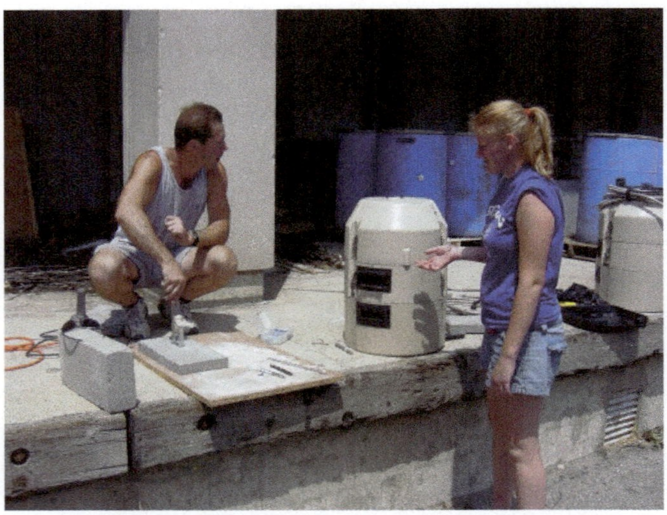

Environmental scientists preparing water autosamplers.

Water pollution may be analyzed through several broad categories of methods: physical, chemical and biological. Most involve collection of samples, followed by specialized analytical tests. Some methods may be conducted *in situ*, without sampling, such as temperature. Government agencies and research organizations have published standardized, validated analytical test methods to facilitate the comparability of results from disparate testing events.

Sampling

Sampling of water for physical or chemical testing can be done by several methods, depending on the accuracy needed and the characteristics of the contaminant. Many contamination events are sharply restricted in time, most commonly in association with rain events. For this reason "grab" samples are often inadequate for fully quantifying contaminant levels. Scientists gathering this type of data often employ auto-sampler devices that pump increments of water at either time or discharge intervals.

Sampling for biological testing involves collection of plants and/or animals from the surface water body. Depending on the type of assessment, the organisms may be identified for biosurveys (population counts) and returned to the water body, or they may be dissected for bioassays to determine toxicity.

Physical Testing

Common physical tests of water include temperature, solids concentrations (e.g., total suspended solids (TSS)) and turbidity.

Chemical Testing

Water samples may be examined using the principles of analytical chemistry. Many published test methods are available for both organic and inorganic compounds. Frequently used methods include pH, biochemical oxygen demand (BOD), chemical oxygen demand (COD), nutrients (nitrate and phosphorus compounds), metals (including copper, zinc, cadmium, lead and mercury), oil and grease, total petroleum hydrocarbons (TPH), and pesticides.

Biological Testing

Biological testing involves the use of plant, animal, and/or microbial indicators to monitor the health of an aquatic ecosystem. They are any biological species or group of species whose function, population, or status can reveal what degree of ecosystem or environmental integrity is present. One example of a group of bio-indicators are the copepods and other small water crustaceans that are present in many water bodies. Such organisms can be monitored for changes (biochemical, physiological, or behavioral) that may indicate a problem within their ecosystem.

Control of Pollution

Decisions on the type and degree of treatment and control of wastes, and the disposal and use of adequately treated wastewater, must be based on a consideration all the technical factors of each drainage basin, in order to prevent any further contamination or harm to the environment.

Sewage Treatment

In urban areas of developed countries, domestic sewage is typically treated by centralized sewage treatment plants. Well-designed and operated systems (i.e., secondary treatment or better) can remove 90 percent or more of the pollutant load in sewage. Some plants have additional systems to remove nutrients and pathogens.

Deer Island Wastewater Treatment Plant serving Boston, Massachusetts and vicinity.

Cities with sanitary sewer overflows or combined sewer overflows employ one or more engineering approaches to reduce discharges of untreated sewage, including:

- utilizing a green infrastructure approach to improve storm water management capacity throughout the system, and reduce the hydraulic overloading of the treatment plant
- repair and replacement of leaking and malfunctioning equipment
- increasing overall hydraulic capacity of the sewage collection system (often a very expensive option).

A household or business not served by a municipal treatment plant may have an individual septic tank, which pre-treats the wastewater on site and infiltrates it into the soil.

Industrial Wastewater Treatment

Dissolved air flotation system for treating industrial wastewater.

Some industrial facilities generate ordinary domestic sewage that can be treated by municipal facilities. Industries that generate wastewater with high concentrations of conventional pollutants (e.g. oil and grease), toxic pollutants (e.g. heavy metals, volatile organic compounds) or other non-conventional pollutants such as ammonia, need specialized treatment systems. Some of these facilities can install a pre-treatment system to remove the toxic components, and then send the partially treated wastewater to the municipal system. Industries generating large volumes of wastewater typically operate their own complete on-site treatment systems. Some industries have been successful at redesigning their manufacturing processes to reduce or eliminate pollutants, through a process called pollution prevention.

Heated water generated by power plants or manufacturing plants may be controlled with:

- cooling ponds, man-made bodies of water designed for cooling by evaporation, convection, and radiation
- cooling towers, which transfer waste heat to the atmosphere through evaporation and/or heat transfer
- cogeneration, a process where waste heat is recycled for domestic and/or industrial heating purposes.

Agricultural Wastewater Treatment

Non Point Source Controls

Riparian buffer lining a creek in Iowa.

Sediment (loose soil) washed off fields is the largest source of agricultural pollution in the United States. Farmers may utilize erosion controls to reduce runoff flows and retain soil on their fields. Common techniques include contour plowing, crop mulching, crop rotation, planting perennial crops and installing riparian buffers.

Nutrients (nitrogen and phosphorus) are typically applied to farmland as commercial fertilizer, animal manure, or spraying of municipal or industrial wastewater (effluent) or sludge. Nutrients may also enter runoff from crop residues, irrigation water, wildlife, and atmospheric deposition. Farmers can develop and implement nutrient management plans to reduce excess application of nutrients and reduce the potential for nutrient pollution.

To minimize pesticide impacts, farmers may use Integrated Pest Management (IPM) techniques (which can include biological pest control) to maintain control over pests, reduce reliance on chemical pesticides, and protect water quality.

Point Source Wastewater Treatment

Feedlot in the United States

Farms with large livestock and poultry operations, such as factory farms, are called *concentrated animal feeding operations* or *feedlots* in the US and are being subject to increasing government regulation. Animal slurries are usually treated by containment in anaerobic lagoons before disposal by spray or trickle application to grassland. Constructed wetlands are sometimes used to facilitate treatment of animal wastes. Some animal slurries are treated by mixing with straw and composted at high temperature to produce a bacteriologically sterile and friable manure for soil improvement.

Erosion and Sediment Control from Construction Sites

Silt fence installed on a construction site.

Sediment from construction sites is managed by installation of:

- erosion controls, such as mulching and hydroseeding, and
- sediment controls, such as sediment basins and silt fences.

Discharge of toxic chemicals such as motor fuels and concrete washout is prevented by use of:

- spill prevention and control plans, and
- specially designed containers (e.g. for concrete washout) and structures such as overflow controls and diversion berms.

Control of Urban Runoff (Storm Water)

Retention basin for controlling urban runoff

Effective control of urban runoff involves reducing the velocity and flow of storm water, as well as reducing pollutant discharges. Local governments use a variety of storm water management techniques to reduce the effects of urban runoff. These techniques, called best management practices (BMPs) in the U.S., may focus on water quantity control, while others focus on improving water quality, and some perform both functions.

Pollution prevention practices include low-impact development techniques, installation of green roofs and improved chemical handling (e.g. management of motor fuels & oil, fertilizers and pesticides). Runoff mitigation systems include infiltration basins, bioretention systems, constructed wetlands, retention basins and similar devices.

Thermal pollution from runoff can be controlled by storm water management facilities that absorb the runoff or direct it into groundwater, such as bioretention systems and infiltration basins. Retention basins tend to be less effective at reducing temperature, as the water may be heated by the sun before being discharged to a receiving stream.

Sources of Sewage

The wastewater generated from the household activities contributes to the major part of the sewage. The wastewater generated from recreational activities, public utilities, commercial complexes, and institutions is also discharged in to sewers. The wastewater discharged from small and medium scale industries situated within the municipal limits and discharging partially treated or untreated wastewater in to the sewers also contributes for municipal wastewater.

Sewage Discharge

The quality of sewage and its characteristics show a marked range of hourly variation and hence peak, average and minimum discharge are important considerations. The process loadings in the sewage treatment are based on the daily average characteristics as determined from a 24 hours weighted composite samples. In the absence of any data an average quantity of 150 LPCD may be adopted for design. The hydraulic design load varies from component to component of the treatment plant with all appurtenances, conduits, channels etc., being designed for the maximum discharge, which may vary from 2.0 to 3.5 times the average discharge. Sedimentation tanks are designed on the basis of average discharge, while consideration of both maximum and minimum discharge is important in the design of screens and grit chambers. Secondary treatment is generally designed for average discharge, with sufficient safety margin to accommodate the peak discharge.

Effect of Untreated Wastewater Disposal

The daily activities of human beings produce both liquid and solid wastes. The liquid portion of the wastewater is necessarily the water supplied by the authority or through private water sources, after it has fouled by variety of uses. The sources of wastewater generation can be defined as a combination of the liquid or water-carried wastes removed from residences, institutions, and commercial and industrial establishments, together with groundwater, surface water, and storm water as may be present. If the untreated wastewater is allowed to accumulate, it will lead to highly unhygienic conditions. The organic matter present in the wastewater will undergo decomposition with production of large quantities of malodorous gases. If the wastewater is discharged without treatment in the water body, this will result in the depletion of Dissolved Oxygen (DO) from the water bodies. Due to depletion of DO, the survival of aquatic life will become difficult, finally leading to anaerobic conditions in the receiving waters. The nutrients present in the wastewater can stimulate the growth of aquatic plants, leading to problems like eutrophication. In addition, the untreated domestic wastewater usually contains numerous pathogenic or disease causing microorganisms, that dwell in the human intestinal tract or it may be present in certain industrial wastewaters. Apart from this, the wastewater contains inorganic gritty materials. The continuous deposition of this inorganic material may reduce the capacity of water body considerably over a period.

Generally domestic sewage does not contain any inorganic matter or organic compounds in highly toxic concentration. However, depending upon the type of industries discharging into the public sewers and the dilution that is offered by sewage; the municipal wastewater may have these inorganic substances or toxic organic compounds with the concentration more than the discharge limits stipulated by the authorities. Certain compounds, such as sulphates, metals such as chromium, etc., if presents in higher concentration, may disturb the secondary treatment of the sewage.

Objectives of Sewage Collection and Disposal

The objective of sewage collection and disposal is to ensure that sewage discharged from communities is properly collected, transported, treated to the required degree so as not to cause danger to human health or unacceptable damage to the natural environment and finally disposed off without causing any health or environmental problems. Thus, efficient sewerage scheme can achieve the following:

- To provide a good sanitary environmental condition of city protecting public health.
- To dispose the human excreta to a safe place by a safe and protective means.
- To dispose of all liquid waste generated from community to a proper place to prevent a favorable condition for mosquito breeding, fly developing or bacteria growing.
- To treat the sewage, as per needs, so as not to endanger the body of water or groundwater or land to get polluted where it is finally disposed off. Thus, it protects the receiving environment from degradation or contamination.

History of Water Supply and Sanitation

Aqueduct in Petra, Jordan.

Water supply and sanitation has been a primary logistical challenge since the dawn of civilization. Where water resources or infrastructure or sanitation systems are insufficient for the population, people fall prey to disease, dehydration, and in extreme cases, death.

Major human settlements could initially develop only where fresh surface water was plentiful, such as near rivers or natural springs. Throughout history people have devised systems to make getting water into their communities and households, and disposing (and later also treating) wastewater more convenient.

The historical focus of sewage treatment was on the conveyance of raw sewage to a natural body of water, e.g. a river or ocean, where it would be satisfactorily diluted and dissipated. Early human habitations were often built next to water sources. Rivers would often double as a crude form of natural sewage disposal.

Most of these early sewers received significant amounts of draft animal dung in street runoff; but handling of human waste varied with location. Over the millennia, technology has dramatically increased the distances across which water can be relocated, but the availability of clean and fresh water remains a limiting factor on the size and density of population centers, and is expected to remain so into the foreseeable future.

Prehistory

Skara Brae a Neolithic village in Orkney, Scotland with home furnishings including water-flushing toilets 3180 BC–2500 BC

During the Neolithic, humans dug the first permanent water wells, from where vessels could be filled and carried by hand. Wells dug around 6500 BC have been found in the Jezreel Valley. The size of human settlements was largely dependent on nearby available water.

A primitive indoor, tree bark lined, two channel, stone, fresh and wastewater system appears to have featured in the houses of in Skara Brae, from around 3000 BCE, along with a cell like enclave in a number of houses, that it has been suggested may have functioned as an early indoor toilet.

Indus Valley Civilization

The Indus Valley Civilization in Asia shows early evidence of public water supply and sanitation. The system the Indus developed and managed included a number of advanced features. A typical example is the Indus city of Lothal (c. 2350 BCE). In Lothal all houses had their own private toilet which was connected to a covered sewer network constructed of brickwork held together with a gypsum-based mortar that emptied either into the surrounding water bodies or alternatively into cesspits, the latter of which were regularly emptied and cleaned.

A large well and bathing platforms at Harappa, remains of the city's final phase of occupation from 2200 to 1900 BC.

The urban areas of the Indus Valley civilization included public and private baths. Sewage was disposed through underground drains built with precisely laid bricks, and a sophisticated water management system with numerous reservoirs was established. In the drainage systems, drains from houses were connected to wider public drains.

Devices such as shadoofs and sakias were used to lift water to ground level. Ruins from the Indus Valley Civilization like Mohenjo-daro in Pakistan and Dholavira in Gujarat in India had settlements with some of the ancient world's most sophisticated sewage systems. They included drainage channels, rain water harvesting, and street ducts.

Stepwells have mainly been used in Indian subcontinent.

The earliest covered sewers uncovered by archaeologists are in the regularly planned cities of the Indus Valley Civilization.

Wastewater Reuse Activities

Wastewater reuse is an ancient practice, which has been applied since the dawn of human history, and is connected to the development of sanitation provision. Reuse of untreated municipal wastewater has been practiced for many centuries with the objective of diverting human waste outside of urban settlements. Likewise, land application of domestic wastewater is an old and common practice, which has gone through different stages of development.

Domestic wastewater was used for irrigation by prehistoric civilizations (e.g. Mesopotamian, Indus valley, and Minoan) since the Bronze Age (ca. 3200-1100 BC). Thereafter, wastewater was used for disposal, irrigation, and fertilization purposes by Hellenic civilizations and later by Romans in areas surrounding cities (e.g. Athens and Rome). Moreover, in China, use of human excreta for fertilizing agricultural crops has been practiced since ancient time.

Bronze Age

Mesopotamia

The Babylonians introduced the world to clay sewer pipes, c4000 BCE, with the earliest examples found in the Temple of Bel at Nippur and at Eshnunna, Babylonia.

The first sanitation systems were built in the prehistoric Middle East, in the south-east of the modern country of Iran near Zabol.

Ancient Greek Civilization

The ancient Greek civilization of Crete, known as the Minoan civilization, was the first civilization to use underground clay pipes for sanitation and water supply. Their capital, Knossos, had a well-organized water system for bringing in clean water, taking out waste water and storm sewage canals for overflow when there was heavy rain. It was also one of the first uses of a flush toilet, dating back to 18th century BC. The Minoan civilization had stone sewers that were periodically flushed with clean water. In addition to sophisticated water and sewer systems they devised elaborate heating systems. The Ancient Greeks of Athens and Asia Minor also used an indoor plumbing system, used for pressurized showers. The Greek inventor Heron used pressurized piping for fire fighting purposes in the City of Alexandria. The Mayans were the third earliest civilization to have employed a system of indoor plumbing using pressurized water.

An inverted siphon system, along with glass covered clay pipes, was used for the first time in the palaces of Crete, Greece. It is still in working condition, after about 3000 years.

Roman Empire

Pont du Gard, a Roman aqueduct in France.

In ancient Rome, the Cloaca Maxima, considered a marvel of engineering, disgorged into the Tiber.

Public latrines were built over the Cloaca Maxima.

The roman empire had indoor plumbing, meaning a system of aqueducts and pipes that terminated in homes and at public wells and fountains for people to use. Rome and other nations used lead pipes; while commonly thought to be the cause of lead poisoning in the Roman Empire, the combination of running water which did not stay in contact with the pipe for long and the deposition of precipitation scale actually mitigated the risk from lead pipes.

Roman towns and garrisons in the United Kingdom between 46 BC and 400 AD had complex sewer networks sometimes constructed out of hollowed-out elm logs, which were shaped so that they butted together with the down-stream pipe providing a socket for the upstream pipe.

Others

Plumbing is also known to have been used in East Asia since the Qin and Han Dynasties of China.

During the Zhou Dynasty in ancient China, sewers existed in various cities such as Linzi.

A Chinese ceramic model of a well with a pulley system, excavated from a tomb of the Han Dynasty (202 BC - 220 AD) period

Middle and Early Modern Age

A 1939 conceptual illustration showing various ways that typhoid bacteria can contaminate a water well (center)

Waterworks (*Wasserkunst*) and fountain from 1602 in Wismar, Germany.

In medieval European cities, small natural waterways used for carrying off wastewater were eventually covered over and functioned as sewers. London's River Fleet is such a system. Open drains, or gutters, for waste water run-off ran along the center of some streets. These were known as "kennels" (i.e., canals, channels), and in Paris were sometimes known as "split streets," as the waste water running along the middle physically split the streets into two halves. The first closed sewer constructed in Paris was designed by HuguesAubird in 1370 on Monmartre Street, and was 300 meters long. The original purpose of designing and constructing a closed sewer in Paris was less-so for waste management as much as it was to hold back the stench coming from the odorous waste water.

Pail closets, outhouses, and cesspits were used to collect human waste. The use of human waste as fertilizer was especially important in China and Japan, where cattle manure was less available. However, most cities did not have a functioning sewer system before the Industrial era, relying instead on nearby rivers or occasional rain showers to wash away the sewage from the streets. In some places, waste water simply ran down the streets, which had stepping stones to keep pedestrians out of the muck, and eventually drained as runoff into the local watershed.

In the 16th century, Sir John Harington invented a flush toilet as a device for Queen Elizabeth I (his godmother) that released wastes into cesspools.

After the adoption of gunpowder, municipal outhouses became an important source of raw material for the making of saltpeter in European countries.

In London, the contents of the city's outhouses were collected every night by commissioned wagons and delivered to the nitrite beds where it was sown into the special soil beds to produce earth rich in mineral nitrates. The nitrate rich-earth is then further processed to produce saltpeter, or potassium nitrate, an important ingredient in black powder.

Classic and Early Modern Mesoamerica

The Classic Maya at Palenque had underground aqueducts and flush toilets; the Classic Maya even used household water filters using locally abundant limestone carved into a porous cylinder, made so as to work in a manner strikingly similar to Modern ceramic water filters.

Sewage farms for Disposal and Irrigation

"Sewage farms" (i.e. wastewater application to the land for disposal and agricultural use) were operated in Bunzlau (Silesia) in 1531, in Edinburgh (Scotland) in 1650, in Paris (France) in 1868, in Berlin (Germany) in 1876 and in different parts of the USA since 1871, where wastewater was used for beneficial crop production. In the following centuries (16th and 18th centuries) in many rapidly growing countries/cities of Europe (e.g. Germany, France) and the United States, "sewage farms" were increasingly seen as a solution for the disposal of large volumes of the wastewater, some of which are still in operation today. Irrigation with sewage and other wastewater effluents has a long history also in China and India; while also a large "sewage farm" was established in Melbourne, Australia in 1897.

Modern Age

Sewer Systems

A significant development was the construction of a network of sewers to collect wastewater. In some cities, including Rome, Istanbul (Constantinople) and Fustat, networked ancient sewer systems continue to function today as collection systems for those cities' modernized sewer systems. Instead of flowing to a river or the sea, the pipes have been re-routed to modern sewer treatment facilities.

Many industrialized cities had incomplete public sanitation well into the 20th century. Outhouses in Brisbane, Australia, around 1950.

However, until the Enlightenment era, little progress was made in water supply and sanitation and the engineering skills of the Romans were largely neglected throughout Europe. This began to change in the 17th and 18th centuries with a rapid expansion in waterworks and pumping systems.

The tremendous growth of cities during the Industrial Revolution quickly led to terribly overpolluted streets, which acted as a constant source for the outbreak of disease.

People wealthy enough to enjoy 19th century flush toilets often had the political power to allow them to drain into public sewers; and the practice became the norm as indoor plumbing became more common.

As cities grew in the 19th century, increasing concerns were raised about public health. As part of a trend of municipal sanitation programs in the late 19th and 20th centuries, many cities constructed extensive sewer systems to help control outbreaks of disease such as typhoid and cholera.

Initially these systems discharged sewage directly to surface waters without treatment. Later, cities attempted to treat the sewage before discharge in order to prevent water pollution and water-borne diseases. During the half-century around 1900, these public health interventions succeeded in drastically reducing the incidence of water-borne diseases among the urban population, and were an important cause in the increases of life expectancy experienced at the time.

Early techniques involved land application of sewage on agricultural land. The use of the land treatment systems continued into the nineteenth/twentieth century in central Europe, USA, and other locations all over the world, but not without causing serious public health concerns and negative environmental impacts. During 1840s and 1850s, this practice resulted in disastrous spread of waterborne diseases like cholera and typhoid. However, when the water supply links with these diseases became clear, engineering solutions were implemented that include the development of alternative water sources using reservoirs and aqueduct systems, relocation of water intakes, and water and wastewater treatment systems.

In the late 19th century some cities began to add chemical treatment and sedimentation systems to their sewers. Most cities in the Western world added more expensive systems for sewage treatment in the 20th century, after scientists at the University of Manchester discovered the sewage treatment process of activated sludge in 1912.

Storm and sanitary sewers were necessarily developed along with the growth of cities. By the 1840s the luxury of indoor plumbing, which mixes human waste with water and flushes it away, eliminated the need for cesspools. Odor was considered the big problem in waste disposal and to address it, sewage could be drained to a lagoon, or "settled" and the solids removed, to be disposed of separately. This process is now called "primary treatment" and the settled solids are called "sludge."

London and other Cities, UK

As recently as the late 19th century sewerage systems in some parts of the rapidly industrializing United Kingdom were so inadequate that water-borne diseases such as cholera and typhoid remained a risk.

From as early as 1535 there were efforts to stop polluting the River Thames in London. Beginning with an Act passed that year that was to prohibit the dumping of excrement into the river. Leading up to the Industrial Revolution the River Thames was identified as being thick and black due to sewage, and it was even said that the river "smells like death." As Britain was the first country to industrialize, it was also the first to experience the disastrous consequences of major urbanisation and was the first to construct a modern sewerage system to mitigate the resultant unsanitary conditions. During the early 19th century, the River Thames was effectively an open sewer, leading to frequent outbreaks of cholera epidemics. Proposals to modernise the sewerage system had been made during 1856, but were neglected due to lack of funds. However, after the *Great Stink* of 1858, Parliamentrealised the urgency of the problem and resolved to create a modern sewerage system.

Joseph Bazalgette, a civil engineer and Chief Engineer of the Metropolitan Board of Works, was given responsibility for the work. He designed an extensive underground sewerage system that diverted waste to the Thames Estuary, downstream of the main centre of population. Six main interceptor sewers, totalling almost 100 miles (160 km) in length, were constructed, some incorporating stretches of London's 'lost' rivers. Three of these sewers were north of the river, the southernmost, low-level one being incorporated in the Thames Embankment. The Embankment also allowed new roads, new public gardens, and the Circle Line of the London Underground.

The intercepting sewers, constructed between 1859 and 1865, were fed by 450 miles (720 km) of main sewers that, in turn, conveyed the contents of some 13,000 miles (21,000 km) of smaller local sewers. Construction of the interceptor system required 318 million bricks, 2.7 million cubic metres of excavated earth and 670,000 cubic metres of concrete.Gravity allowed the sewage to flow eastwards, but in places such as Chelsea, Deptford and Abbey Mills, pumping stations were built to raise the water and provide sufficient flow. Sewers north of the Thames feed into the Northern Outfall Sewer, which fed into a major treatment works at Beckton. South of the river, the Southern Outfall Sewer extended to a similar facility at Crossness. With only minor modifications, Bazalgette's engineering achievement remains the basis for sewerage design up into the present day.

In MerthyrTydfil, a large town in South Wales, most houses discharged their sewage to individual cess-pits which persistently overflowed causing the pavements to be awash with foul sewage.

Paris, France

In 1802 Napoleon built the Ourcq canal which brought 70,000 cubic meters of water a day to Paris, while the Seine river received up to 100,000 cubic meters of wastewater per day. The Paris cholera epidemic of 1832 sharpened the public awareness of the necessity for some sort of drainage system to deal with sewage and waste water in a better and healthier way. Between 1865 and 1920 Eugene Belgrand lead the development of a large scale system for water supply and waste water management. Between these years approximately 600 kilometres of aqueducts were built to bring in potable spring water, which freed the poor quality water to be used for flushing streets and sewers. By 1894 laws were passed which made drainage mandatory. The treatment of Paris sewage, though, was left to natural devices as 5,000 hectares of land were used to spread the waste out to be naturally purified. Further, the lack of sewage treatment left Parisian sewage pollution to become concentrated downstream in the town of Clichy, effectively forcing residents to pack up and move elsewhere.

The 19th century brick-vaulted Paris sewers serve as a tourist attraction nowadays.

Hamburg and Frankfurt, Germany

The first comprehensive sewer system in a German city was built in Hamburg, Germany in the mid-19th century.

In 1863, work began on the construction of a modern sewerage system for the rapidly growing city of Frankfurt am Main, based on design work by William Lindley. 20 years after the system's completion, the death rate from typhoid had fallen from 80 to 10 per 100,000 inhabitants.

USA

The first sewer systems in the United States were built in the late 1850s in Chicago and Brooklyn.

In the United States, the first sewage treatment plant using chemical precipitation was built in Worcester, Massachusetts in 1890.

Sewage Treatment Plants

At the end of the 19th century, since primary treatment still left odor problems, it was discovered that bad odors could be prevented by introducing oxygen into the decomposing sewage. This was the beginning of the biological aerobic and anaerobic treatments which are fundamental to waste water processes.

As pollution of water bodies became a concern, cities attempted to treat the sewage before discharge. In the late 19th century some cities began to add chemical treatment and sedimentation systems to their sewers. In the United States, the first sewage treatment plant using chemical precipitationwas built in Worcester, Massachusetts in 1890. Most cities in the Western world added more expensive systems for sewage treatment in the early 20th century, after scientists at the University of Manchester discovered the sewage treatment process of activated sludge in 1912. During the half-century around 1900, these public health interventions succeeded in drastically reducing the incidence of water-borne diseases among the urban population, and were an important cause in the increases of life expectancy experienced at the time.

Toilets

With the onset of the industrial revolution and related advances in technology, the flush toilet began to emerge into its modern form. It needs to be connected to a sewer system though. Where this is not feasible or desired, dry toilets are an alternative option.

Water Supply

Chelsea Waterworks, 1752. Two Newcomen beam engines pumped Thames water from a canal to reservoirs at Green Park and Hyde Park.

An ambitious engineering project to bring fresh water from Hertfordshire to London was undertaken by Hugh Myddleton who oversaw the construction of the New River between 1609 and 1613. The New River Company became one of the largest private water companies of the time, supplying the City of London and other central areas.

It was in the 18th century that a rapidly growing population fueled a boom in the establishment of private water supply networks in London. The Chelsea Waterworks Company was established in 1723 "for the better supplying the City and Liberties of Westminster and parts adjacent with water". The company created extensive ponds in the area bordering Chelsea and Pimlico using water from the tidal Thames. Other waterworks were established in London, including at West Ham in 1743, at Lea Bridge before 1767, Lambeth Waterworks Company in 1785, West Middlesex Waterworks Company in 1806 and Grand Junction Waterworks Company in 1811.

The S-bendpipe was invented by Alexander Cummings in 1775 but became known as the U-bend following the introduction of the U-shaped trap by Thomas Crapper in 1880. The first screw-down water tap was patented in 1845 by Guest and Chrimes, a brass foundry in Rotherham.

Water Treatment

Founders of microscopy, Antonie van Leeuwenhoek and Robert Hooke, used the newly invented microscope to observe for the first time small material particles that were suspended in the water, laying the groundwork for the future understanding of waterborne pathogens and waterborne diseases.

Sand Filter

Original map by John Snow showing the clusters of cholera cases in the London epidemic of 1854.

Sir Francis Bacon attempted to desalinate sea water by passing the flow through a sand filter. Although his experiment didn't succeed, it marked the beginning of a new interest in the field.

The first documented use of sand filters to purify the water supply dates to 1804, when the owner of a bleachery in Paisley, Scotland, John Gibb, installed an experimental filter, selling his unwanted surplus to the public. This method was refined in the following two decades by engineers working for private water companies, and it culminated in the first treated public water supply in the world, installed by engineer James Simpson for the Chelsea Waterworks Company in London in 1829. This installation provided filtered water for every resident of the area, and the network design was widely copied throughout the United Kingdom in the ensuing decades.

The practice of water treatment soon became mainstream, and the virtues of the system were made starkly apparent after the investigations of the physician John Snow during the 1854 Broad Street cholera outbreak. Snow was sceptical of the then-dominant miasma theory that stated that diseases were caused by noxious "bad airs". Although the germ theory of disease had not yet been developed, Snow's observations led him to discount the prevailing theory. His 1855 essay *On the Mode of Communication of Cholera* conclusively demonstrated the role of the water supply in spreading the cholera epidemic in Soho, with the use of a dot distribution map and statistical proof to illustrate the connection between the quality of the water source and cholera cases. His data convinced the local council to disable the water pump, which promptly ended the outbreak.

The Metropolis Water Act introduced the regulation of the water supply companies in London, including minimum standards of water quality for the first time. The Act "made provision for securing the supply to the Metropolis of pure and wholesome water", and required that all water be "effectually filtered" from 31 December 1855. This was followed up with legislation for the mandatory inspection of water quality, including comprehensive chemical analyses, in 1858. This legislation set a worldwide precedent for similar state public health interventions across Europe. The Metropolitan Commission of Sewers was formed at the same time, water filtration was adopted throughout the country, and new water intakes on the Thames were established above Teddington Lock. Automatic pressure filters, where the water is forced under pressure through the filtration system, were innovated in 1899 in England.

Water Chlorination

In what may have been one of the first attempts to use chlorine, William Soper used chlorinated lime to treat the sewage produced by typhoid patients in 1879.

In a paper published in 1894, Moritz Traube formally proposed the addition of chloride of lime (calcium hypochlorite) to water to render it "germ-free." Two other investigators confirmed Traube's findings and published their papers in 1895. Early attempts at implementing water chlorination at a water treatment plant were made in 1893 in Hamburg, Germany and in 1897 the city of MaidstoneEngland was the first to have its entire water supply treated with chlorine.

Permanent water chlorination began in 1905, when a faulty slow sand filter and a contaminated water supply led to a serious typhoid fever epidemic in Lincoln, England. Dr. Alexander Cruickshank Houston used chlorination of the water to stem the epidemic. His installation fed a concentrated solution of chloride of lime to the water being treated. The chlorination of the water supply helped stop the epidemic and as a precaution, the chlorination was continued until 1911 when a new water supply was instituted.

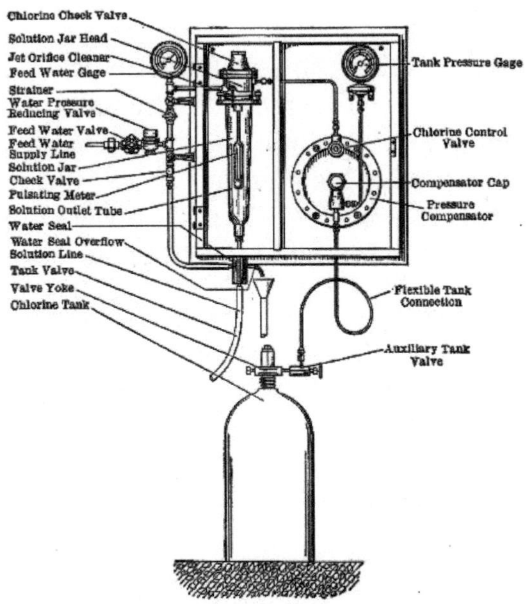

Manual Control Chlorinator for the liquefaction of chlorine for water purification, early 20th century.
From *Chlorination of Water* by Joseph Race, 1918.

The first continuous use of chlorine in the United States for disinfection took place in 1908 at Boonton Reservoir (on the Rockaway River), which served as the supply for Jersey City, New Jersey. Chlorination was achieved by controlled additions of dilute solutions of chloride of lime (calcium hypochlorite) at doses of 0.2 to 0.35 ppm. The treatment process was conceived by Dr.

John L. Leal and the chlorination plant was designed by George Warren Fuller. Over the next few years, chlorine disinfection using chloride of lime were rapidly installed in drinking water systems around the world.

The technique of purification of drinking water by use of compressed liquefied chlorine gas was developed by a British officer in the Indian Medical Service, Vincent B. Nesfield, in 1903. According to his own account, "It occurred to me that chlorine gas might be found satisfactory ... if suitable means could be found for using it.... The next important question was how to render the gas portable. This might be accomplished in two ways: By liquefying it, and storing it in lead-lined iron vessels, having a jet with a very fine capillary canal, and fitted with a tap or a screw cap. The tap is turned on, and the cylinder placed in the amount of water required. The chlorine bubbles out, and in ten to fifteen minutes the water is absolutely safe. This method would be of use on a large scale, as for service water carts."

U.S. ArmyMajorCarl Rogers Darnall, Professor of Chemistry at the Army Medical School, gave the first practical demonstration of this in 1910. Shortly thereafter, Major William J. L. Lyster of the Army Medical Department used a solution of calcium hypochlorite in a linen bag to treat water. For many decades, Lyster's method remained the standard for U.S. ground forces in the field and in camps, implemented in the form of the familiar Lyster Bag (also spelled Lister Bag). This work became the basis for present day systems of municipal water *purification*.

Fluoridation

Water fluoridation is a controversial practice has been carried out since the early 20th century for the claimed purpose of decreasing tooth decay.

Today (21st Century and Beyond)

Decentralization of water infrastructure has grown extensively as a viable solution including rainwater harvesting and stormwater harvesting where policies are eventually tending towards a more rational use and sourcing of water incorporation concepts such as "Fit for Purpose".

The Sustainable Development Goals formulated in 2015 include targets on access to water supply and sanitation at a global level.

During the beginning of the 21st century, especially in areas of urban and suburban population centres, traditional centralized infrastructure have not been able to supply sufficient quantities of water to keep up with growing demand. One option is the use of desalinationtechnology, this is especially prevalent in coastal areas and in "dry" countries like Australia.

Wastewater

Wastewater, also written as waste water, is any water that has been adversely affected in quality by anthropogenic influence. Wastewater can originate from a combination of domestic, industrial, commercial or agricultural activities, surface runoff or stormwater, and from sewer inflow or infiltration.

Municipal wastewater (also called sewage) is usually conveyed in a combined sewer or sanitary sewer, and treated at a wastewater treatment plant. Treated wastewater is discharged into receiving water via an effluent pipe. Wastewaters generated in areas without access to centralized sewer systems rely on on-site wastewater systems. These typically comprise a septic tank, drain field, and optionally an on-site treatment unit. The management of wastewater belongs to the overarching term sanitation, just like the management of human excreta, solid waste and stormwater (drainage).

Greywater (a type of wastewater) in a settling tank

Sewage is a type of wastewater that comprises domestic wastewater and is therefore contaminated with feces or urine from people's toilets, but the term sewage is also used to mean any type of wastewater. Sewerage is the physical infrastructure, including pipes, pumps, screens, channels etc. used to convey sewage from its origin to the point of eventual treatment or disposal.

Sources

Wastewater can come from:

- Human excreta (feces and urine) often mixed with used toilet paper or wipes; this is known as blackwater if it is collected with flush toilets
- Washing water (personal, clothes, floors, dishes, cars, etc.), also known as greywater or sullage
- Surplus manufactured liquids from domestic sources (drinks, cooking oil, pesticides, lubricating oil, paint, cleaning liquids, etc.)
- Urban rainfall runoff from roads, carparks, roofs, sidewalks/pavements (contains oils, animal feces, litter, gasoline/petrol, diesel or rubber residues from tires, soapscum, metals from vehicle exhausts, etc.)

- Highway drainage (oil, de-icing agents, rubber residues, particularly from tires)
- Storm drains (may include trash)
- Manmade liquids (illegal disposal of pesticides, used oils, etc.)
- Industrial waste
- Industrial site drainage (silt, sand, alkali, oil, chemical residues);
    - Industrial cooling waters (biocides, heat, slimes, silt)
    - Industrial process waters
    - Organic or biodegradable waste, including waste from abattoirs, creameries, and ice cream manufacture
    - Organic or non bio-degradable/difficult-to-treat waste (pharmaceutical or pesticide manufacturing)
    - Extreme pH waste (from acid/alkali manufacturing, metal plating)
    - Toxic waste (metal plating, cyanide production, pesticide manufacturing, etc.)
    - Solids and emulsions (paper manufacturing, foodstuffs, lubricating and hydraulic oil manufacturing, etc.)
    - Agricultural drainage, direct and diffuse
    - Hydraulic fracturing
    - Produced water from oil & natural gas production

Wastewater can be diluted or mixed with other types of water in the form of:

- Seawater ingress (high volumes of salt and microbes)
- Direct ingress of river water
- Rainfall collected on roofs, yards, hard-standings, etc. (generally clean with traces of oils and fuel)
- Groundwater infiltrated into sewage

After it has undergone some treatment, the “treated wastewater” remains, e.g.:

- Septic tank discharge
- Sewage treatment plant discharge

Constituents

The composition of wastewater varies widely. This is a partial list of what it may contain:

- Water (more than 95 percent), which is often added during flushing to carry waste down a drain;
- Pathogens such as bacteria, viruses, prions and parasitic worms;
- Non-pathogenic bacteria;
- Organic particles such as feces, hairs, food, vomit, paper fibers, plant material, humus, etc.;
- Soluble organic material such as urea, fruit sugars, soluble proteins, drugs, pharmaceuticals, etc.;
- Inorganic particles such as sand, grit, metal particles, ceramics, etc.;
- Soluble inorganic material such as ammonia, road-salt, sea-salt, cyanide, hydrogen sulfide, thiocyanates, thiosulfates, etc.;
- Animals such as protozoa, insects, arthropods, small fish, etc.;
- Macro-solids such as sanitary napkins, nappies/diapers, condoms, needles, children's toys, dead animals or plants, etc.;
- Gases such as hydrogen sulfide, carbon dioxide, methane, etc.;
- Emulsions such as paints, adhesives, mayonnaise, hair colorants, emulsified oils, etc.;
- Toxins such as pesticides, poisons, herbicides, etc.
- Pharmaceuticals and hormones and other hazardous substances

Quality Indicators

Any oxidizable material present in an aerobic natural waterway or in an industrial wastewater will be oxidized both by biochemical (bacterial) or chemical processes. The result is that the oxygen content of the water will be decreased. Basically, the reaction for biochemical oxidation may be written as:

Oxidizable material + bacteria + nutrient + $O_2 \rightarrow CO_2 + H_2O$ + oxidized inorganics such as NO_3^- or SO_4^{2-}

Oxygen consumption by reducing chemicals such as sulfides and nitrites is typified as follows:

$$S^{2-} + 2\,O_2 \rightarrow SO_4^{2-}$$

$$NO_2^- + ½\,O_2 \rightarrow NO_3^-$$

Since all natural waterways contain bacteria and nutrients, almost any waste compounds introduced into such waterways will initiate biochemical reactions (such as shown above). Those biochemical reactions create what is measured in the laboratory as the biochemical oxygen demand (BOD). Such chemicals are also liable to be broken down using strong oxidizing agents and these chemical reactions create what is measured in the laboratory as the chemical oxygen demand (COD). Both the BOD and COD tests are a measure of the relative oxygen-depletion effect of a waste contaminant. Both have been widely adopted as a measure of pollution effect. The BOD test measures the oxygen demand of biodegradable pollutants whereas the COD test measures the oxygen demand of oxidizable pollutants.

The so-called 5-day BOD measures the amount of oxygen consumed by biochemical oxidation of waste contaminants in a 5-day period. The total amount of oxygen consumed when the biochemical reaction is allowed to proceed to completion is called the Ultimate BOD. Because the Ultimate BOD is so time consuming, the 5-day BOD has been almost universally adopted as a measure of relative pollution effect.

There are also many different COD tests of which the 4-hour COD is probably the most common.

There is no generalized correlation between the 5-day BOD and the ultimate BOD. Similarly there is no generalized correlation between BOD and COD. It is possible to develop such correlations for specific waste contaminants in a specific wastewater stream but such correlations cannot be generalized for use with any other waste contaminants or wastewater streams. This is because the composition of any wastewater stream is different. As an example an effluent consisting of a solution of simple sugars that might discharge from a confectionery factory is likely to have organic components that degrade very quickly. In such a case, the 5 day BOD and the ultimate BOD would be very similar since there would be very little organic material left after 5 days. However a final effluent of a sewage treatment works serving a large industrialised area might have a discharge where the ultimate BOD was much greater than the 5 day BOD because much of the easily degraded material would have been removed in the sewage treatment process and many industrial processes discharge difficult to degrade organic molecules.

The laboratory test procedures for the determining the above oxygen demands are detailed in many standard texts. American versions include the "Standard Methods for the Examination of Water and Wastewater."

Treatment

There are numerous processes that can be used to clean up wastewaters depending on the type and extent of contamination. Wastewater can be treated in wastewater treatment plants which include physical, chemical and biological treatment processes. Municipal wastewater is treated in sewage treatment plants (which may also be referred to as wastewater treatment plants). Agricultural wastewater may be treated in agricultural wastewater treatment processes, whereas industrial wastewater is treated in industrial wastewater treatment processes.

For municipal wastewater the use of septic tanks and other On-Site Sewage Facilities (OSSF) is widespread in some rural areas, for example serving up to 20 percent of the homes in the U.S.

One type of aerobic treatment system is the activated sludge process, based on the maintenance and recirculation of a complex biomass composed of micro-organisms able to absorb and adsorb the organic matter carried in the wastewater. Anaerobic wastewater treatment processes (UASB, EGSB) are also widely applied in the treatment of industrial wastewaters and biological sludge. Some wastewater may be highly treated and reused as reclaimed water. Constructed wetlands are also being used.

Disposal

Industrial wastewater effluent with neutralized pH from tailing runoff in Peru.

In some urban areas, municipal wastewater is carried separately in sanitary sewers and runoff from streets is carried in storm drains. Access to either of these is typically through a manhole. During high precipitation periods a combined sewer overflow can occur, forcing untreated sewage to flow back into the environment. This can pose a serious threat to public health and the surrounding environment.

Sewage may drain directly into major watersheds with minimal or no treatment but this usually has serious impacts on the quality of an environment and on the health of people. Pathogens can cause a variety of illnesses. Some chemicals pose risks even at very low concentrations and can remain a threat for long periods of time because of bioaccumulation in animal or human tissue.

Wastewater may be pumped underground through an injection well.

Reuse

Treated wastewater can be reused as drinking water, in industry (for example in cooling towers), in artificial recharge of aquifers, in agriculture and in the rehabilitation of natural ecosystems (for example in Florida's Everglades).

Gardening and Agriculture

There are benefits of using recycled water for irrigation, including the lower cost compared to some other sources and consistency of supply regardless of season, climatic conditions and associated water restrictions. Irrigation with recycled wastewater can also serve to fertilize plants if it contains nutrients, such as nitrogen, phosphorus and potassium.

Around 90% of wastewater produced globally remains untreated, causing widespread water pollution, especially in low-income countries. Increasingly, agriculture is using untreated wastewater for irrigation. Cities provide lucrative markets for fresh produce, so are attractive to farmers. However, because agriculture has to compete for increasingly scarce water resources with industry and municipal users, there is often no alternative for farmers but to use water polluted with urban waste directly to water their crops.

Health Risks

There can be significant health hazards related to using untreated wastewater in agriculture. Wastewater from cities can contain a mixture of chemical and biological pollutants. In low-income countries, there are often high levels of pathogens from excreta, while in emerging nations, where industrial development is outpacing environmental regulation, there are increasing risks from inorganic and organic chemicals. The World Health Organization, in collaboration with the Food and Agriculture Organization of the United Nations (FAO) and the United Nations Environmental Program (UNEP), has developed guidelines for safe use of wastewater in 2006. These guidelines advocate a 'multiple-barrier' approach to wastewater use, for example by encouraging farmers to adopt various risk-reducing behaviours. These include ceasing irrigation a few days before harvesting to allow pathogens to die off in the sunlight, applying water carefully so it does not contaminate leaves likely to be eaten raw, cleaning vegetables with disinfectant or allowing fecal sludge used in farming to dry before being used as a human manure.

Legislation

European Union

Council Directive 91/271/EEC on Urban Wastewater Treatment was adopted on 21 May 1991, amended by the Commission Directive 98/15/EC. Commission Decision 93/481/EEC defines the information that Member States should provide the Commission on the state of implementation of the Directive.

United States

The Clean Water Act is the primary federal law in the United States governing water pollution.

Philippines

In the Philippines, Republic Act 9275, otherwise known as the Philippine Clean Water Act of 2004, is the governing law on wastewater management. It states that it is the country's policy to protect, preserve and revive the quality of our fresh, brackish and marine waters, for which wastewater management plays a particular role.

Wastewater Treatment

The treatment and safe disposal of wastewater is necessary. This will facilitate protection of environment and environmental conservation, because the wastewater collected from cities and towns must ultimately be returned to receiving water body or to the land or reused to fulfill certain needs. The sewage treatment plants constructed near the end of nineteenth century were designed to remove suspended matter alone by the principal of simple gravity settling. It soon became apparent that primary treatment alone was insufficient to protect the water quality of the receiving water body. This was mainly due to the presence of organic material, in colloidal and dissolved form, in the sewage after settling. Thus, in the beginning of twentieth century several treatment systems, called secondary treatment, were developed with the objective of organic matter removal. For this secondary treatment, biological methods aregenerally used. The aerobic biological treatment processes were popularly used as a secondary treatment, and these processes are still at the first choice.

In the second half of twentieth century, it became clear that the discharge of effluents from even the most efficient secondary treatment plant could lead to the deterioration of the quality of receiving water body. This could be attributed partly to the discharge of ammonia in the effluent. In the receiving water body this discharge exerts an oxygen demand for the biological oxidation of ammonia to nitrate, a process called nitrification. However, even when nitrification is carried out at the treatment plant itself, the discharge of effluent can still be detrimental to the water quality due to introduction of nitrogen in the form of nitrate and phosphorus as phosphate. The tolerance limits of nitrates for the water when used as raw water for public water supplies and bathing ghats is 50 mg/L as NO_3. The availability of nitrogen and phosphorous tends to cause an excessive growth of aquatic life notably, autotrophic organisms such as algae, that can use carbon dioxide rather than organic material as a sources for cell synthesis. Thus, explosive development of biomass may occur when nitrogen and phosphorus are abundantly available. Although, this biomass may produce photosynthetic oxygen in the water during daytime, after sunset it will consume oxygen, so that the dissolved oxygen concentration will decrease and may reach to the levels that are too low to sustain the life of other (macro) organisms. This phenomenon of eutrophication has led to the development of tertiary treatment systems. In these, nitrogen and/or phosphorus are removed, along with solids and organic materials.

Once the minimum effluent quality has been specified, for maximum allowable concentrations of solids (both suspended and dissolved), organic matter, nutrients, and pathogens, the objective of the treatment is to attain reliably the set standards. The role of design engineer is to develop a treatment scheme that will guarantee the technical feasibility of the scheme, taking into consideration other factors such as construction and maintenance costs, the availability of construction materials and equipment, as well as specialized skilled personals for operation and maintenance of the treatment plant.

Primary treatment consists of screens (for removal of floating matter), grit chamber (for removal of inorganic suspended solids) and primary sedimentation tank (for removal residual settleable solids which are mostly organic). Skimming tanks may be used for removal of oils; however, in conventional treatment plant no separate skimming tank is used and oil removal isachieved by collecting the scum in primary sedimentation tank. This primary treatment alone will not produce an effluent with an acceptable residual organic material concentration. Almost invariably biological methods are used in the treatment systems to effect secondary treatment for removal of organic material. In biological treatment systems, the organic material is metabolized by bacteria. Depending upon the requirement for the final effluent quality, tertiary treatment methods and/or pathogen removal may be included.

Today majority of wastewater treatment plants uses aerobic metabolism for the removal of organic matter. The popularly used aerobic processes are the activated sludge process, oxidation ditch, trickling filter, and aerated lagoons. Stabilization ponds use both the aerobic and anaerobic mechanisms. In the recent years, due to increase in power cost and subsequent increase in operation cost of aerobic processes, more attention is being paid for the use of anaerobic treatment systems for the treatment of wastewater including sewage. Recently the high anaerobic process such as Upflow Anaerobic Sludge Blanket (UASB) reactor is used for sewage treatment at many places.

Depending on the mode of disposal the tertiary treatment may be given for killing pathogens, nutrient removal, suspended solids removal, etc. Generally secondary treatment followed by disinfection will meet the effluent standards for disposal into water bodies. When the treated sewage is disposed off on land for irrigation, the level of disinfection needs will depend on the type of secondary treatment and type of crops with restricted or unrestricted public access.

Important Factors Considered for Selecting Material for Sewer

Following factors should be considered before selecting material for manufacturing sewer pipes:

a. Resistance to Corrosion

Sewer carries wastewater that releases gases such as H_2S. This gas in contact with moisture can be converted into sulfuric acid. The formation of acids can lead to the corrosion of sewer pipe. Hence, selection of corrosion resistance material is must for long life of pipe.

b. Resistance to Abrasion

Sewage contain considerable amount of suspended solids, part of which are inorganic solids such as sand or grit. These particles moving at high velocity can cause wear and tear of sewer pipe internally. This abrasion can reduce thickness of pipe and reduces hydraulic efficiency of the sewer by making the interior surface rough.

c. Strength and Durability

The sewer pipe should have sufficient strength to withstand all the forces that are likely to come on them. Sewers are subjected to considerable external loads of backfill material and traffic load, if any. They are not subjected to internal pressure of water. To withstand external load safely without failure, sufficient wall thickness of pipe or reinforcement is essential. In addition, the material selected should be durable and should have sufficient resistance against natural weathering action to provide longer life to the pipe.

d. Weight of the Material

The material selected for sewer should have less specific weight, which will make pipe light in weight. The lightweight pipes are easy for handling and transport.

e. Imperviousness

To eliminate chances of sewage seepage from sewer to surrounding, the material selected for pipe should be impervious.

f. Economy and Cost

Sewer should be less costly to make the sewerage scheme economical.

g. Hydraulically Efficient

The sewer shall have smooth interior surface to have less frictional coefficient.

Materials for Sewers

Asbestos Cement Sewers

- These are manufactured from a mixture of asbestos fibers, silica and cement. Asbestos fibers are thoroughly mixed with cement to act as reinforcement.
- These pipes are available in size 10 to 100 cm internal diameter and length up to 4.0 m.
- These pipes can be easily assembled without skilled labour with the help of special coupling, called 'Ring Tie Coupling' or Simplex joint.
- The pipe and joints are resistant to corrosion and the joints are flexible to permit 120 deflection for curved laying.
- These pipes are used for vertical transport of water. For example, transport of rainwater from roofs in multistoried buildings, for transport of sewage to grounds, and for transport of less foul sullage i.e., wastewater from kitchen and bathroom.

Advantages

- These pipes are light in weight and hence, easy to carry and transport.
- Easy to cut and assemble without skilled labour.
- Interior is smooth (Manning's n = 0.011) hence, can make excellent hydraulically efficient sewer.

Disadvantages

- These pipes are structurally not very strong.
- These are susceptible to corrosion by sulphuric acid. When bacteria produce H_2S, in presence of water, H_2SO_4 can be formed leading to corrosion of pipe material.

Plain Cement Concrete or Reinforced Cement Concrete

Plain cement concrete (1: 1.5: 3) pipes are available up to 0.45 m diameter and reinforcement cement pipes are available up to 1.8 m diameter. These pipes can be cast in situ or precast pipes. Precast pipes are better in quality than the cast in situ pipes. The reinforcement in these pipes can be different such as single cage reinforced pipes, used for internal pressure less than 0.8 m; double cage reinforced pipes used for both internal and external pressure greater than 0.8 m; elliptical cage reinforced pipes used for larger diameter sewers subjected to external pressure; and Hume pipes with steel shells coated with concrete from inside and outside. Nominal longitudinal reinforcement of 0.25% is provided in these pipes.

Advantages of Concrete Pipes

- Strong in tension as well as compression.
- Resistant to erosion and abrasion.
- They can be made of any desired strength.
- Easily molded, and can be in situ or precast pipes.
- Economical for medium and large sizes.
- These pipes are available in wide range of size and the trench can be opened and backfilled rapidly during maintenance of sewers.

Disadvantages

- These pipes can get corroded and pitted by the action of H_2SO_4.
- The carrying capacity of the pipe reduces with time because of corrosion.
- The pipes are susceptible to erosion by sewage containing silt and grit.

The concrete sewers can be protected internally by vitrified clay linings. With protection lining they are used for almost all the branch and main sewers. Only high alumina cement concrete should be used when pipes are exposed to corrosive liquid like sewage.

Vitrified Clay or Stoneware Sewers

These pipes are used for house connections as well as lateral sewers. The size of the pipe available is 5 cm to 30 cm internal diameter with length 0.9 to 1.2 m. These pipes are rarely manufactured for diameter greater than 90 cm. These are joined by bell and spigot flexible compression joints.

Advantages

- Resistant to corrosion, hence fit for carrying polluted water such as sewage.
- Interior surface is smooth and is hydraulically efficient.
- The pipes are highly impervious.
- Strong in compression.
- These pipes are durable and economical for small diameters.
- The pipe material does not absorb water more than 5% of their own weight, when immersed in water for 24 h.

Disadvantages

- Heavy, bulky and brittle and hence, difficult to transport.

- These pipes cannot be used as pressure pipes, because they are weak in tension.
- These require large number of joints as the individual pipe length is small.

Brick Sewers

This material is used for construction of large size combined sewer or particularly for storm water drains. The pipes are plastered from outside to avoid entry of tree roots and groundwater through brick joints. These are lined from inside with stone ware or ceramic block to make them smooth and hydraulically efficient. Lining also makes the pipe resistant to corrosion.

Cast Iron Sewers

These pipes are stronger and capable to withstand greater tensile, compressive, as well as bending stresses. However, these are costly. Cast iron pipes are used for outfall sewers, rising mains of pumping stations, and inverted siphons, where pipes are running under pressure. These are also suitable for sewers under heavy traffic load, such as sewers below railways and highways. They are used for carried over piers in case of low lying areas. They form 100% leak proof sewer line to avoid groundwater contamination. They are less resistant to corrosion; hence, generally lined from inside with cement concrete, coal tar paint, epoxy, etc. These are joined together by bell and spigot joint. IS:1536-1989 and IS:1537-1976 provides the specifications for spun and vertically cast pipes, respectively.

Steel Pipes

These are used under the situations such as pressure main sewers, under water crossing, bridge crossing, necessary connections for pumping stations, laying pipes over self supporting spans, railway crossings, etc. They can withstand internal pressure, impact load and vibrations much better than CI pipes. They are more ductile and can withstand water hammer pressure better. These pipes cannot withstand high external load and these pipes may collapse when negativepressure is developed in pipes. They are susceptible to corrosion and are not generally used for partially flowing sewers. They are protected internally and externally against the action of corrosion.

Ductile Iron Pipes

Ductile iron pipes can also be used for conveying the sewers. They demonstrate higher capacity to withstand water hammer. The specifications for DI pipes is provided in IS:12288-1987. The predominant wall material is ductile iron, a spheroidized graphite cast iron. Internally these pipes are coated with cement mortar lining or any other polyethylene or poly wrap or plastic bagging/ sleeve lining to inhibit corrosion from the wastewater being conveyed, and various types of external coating are used to inhibit corrosion from the environment. Ductile iron has proven to be a better pipe material than cast iron but they are costly. Ductile iron is still believed to be stronger and more fracture resistant material. However, like most ferrous materials it is susceptible to corrosion. A typical life expectancy of thicker walled pipe could be up to 75 years, however with the current thinner walled ductile pipe the life could be about 20 years in highly corrosive soils without a corrosion control program like cathodic protection.

Plastic Sewers (PVC Pipes)

Plastic is recent material used for sewer pipes. These are used for internal drainage works in house. These are available in sizes 75 to 315 mm external diameter and used in drainage works. They offer smooth internal surface. The additional advantages they offer are resistant to corrosion, light weight of pipe, economical in laying, jointing and maintenance, the pipe is tough and rigid, and ease in fabrication and transport of these pipes.

High Density Polythylene (HDPE)

Pipes Use of these pipes for sewers is recent development. They are not brittle like AC pipes and other pipes and hence hard fall during loading, unloading and handling do not cause any damage to the pipes. They can be joined by welding or can be jointed with detachable joints up to 630 mm diameter (IS:4984-1987). These are commonly used for conveyance of industrial wastewater. They offer all the advantages offered by PVC pipes. PVC pipes offer very little flexibility and normally considered rigid; whereas, HDPE pipes are flexible hence best suited for laying in hilly and uneven terrain. Flexibility allows simple handling and installation of HDPE pipes. Because of low density, these pipes are very light in weight. Due to light in weight, they are easy for handling, this reduces transportation and installation cost. HDPE pipes are non corrosive and offer very smooth inside surface due to which pressure losses are minimal and also this material resist scale formation.

Glass Fiber Reinforced Plastic Pipes

This martial is widely used where corrosion resistant pipes are required. Glass fiber reinforced plastic (GRP) can be used as a lining material for conventional pipes to protect from internal or external corrosion. It is made from the composite matrix of glass fiber, polyester resin and fillers. These pipes have better strength, durability, high tensile strength, low density and high corrosion resistance. These are manufactured up to 2.4 m diameter and up to 18 m length (IS:12709-1989). Glass reinforced plastic pipes represent the ideal solution for transport of any kind of water, chemicals, effluent and sewage, because they combine the advantages of corrosion resistance with a mechanical strength which can be compared with the steel pipes. Typical properties that result in advantages in GRP pipes application can be summarized as follows:

- Light weight of pipes that allows for the use of light laying and transport means.
- Possibility of nesting of different diameters of pipe thus allowing additional saving in transport cost.
- Length of pipe is larger than other pipe materials.
- Easy installation procedures due to the kind of mechanical bell and spigot joint.
- Corrosion resistance material, hence no protections such as coating, painting or cathodic are then necessary.
- Smoothness of the internal wall that minimizes the head loss and avoids the formation of deposits.

- High mechanical resistance due to the glass reinforcement.
- Absolute impermeability of pipes and joints both from external to internal and viceversa.
- Very long life of the material.

Lead Sewers

- They are smooth, soft and can take odd shapes.
- This pipe has an ability to resist sulphide corrosion.
- However, these pipes are very costly.
- These are used in house connection.

Shapes of Sewer Pipes

Sewers are generally circular pipes laid below ground level, slopping continuously towards the outfall. These are designed to flow under gravity. Shapes other than circular are also used.

Other shapes used for sewers are:

a. Standard Egg-shaped sewer

b. New egg-shaped sewer

c. Horse shoe shaped sewer

d. Parabolic shaped sewer

e. Semi-elliptical section

f. Rectangular shape section

g. U-shaped section

h. Semi-circular shaped sewer

i. Basket handled shape sewer

Standard egg-shaped sewers, also called as ovoid shaped sewer, and new or modified egg-shaped sewers are used in combined sewers. These sewers can generate self cleansing velocity during dry weather flow. Horse shoe shaped sewers and semi-circular sections are used for large sewers with heavy discharge such as trunk and outfall sewers. Rectangular or trapezoidal section is used for conveying storm water. U-shaped section is used for larger sewers and especially in open cuts. Other sections of the sewers have become absolute due to difficulty in construction on site and non availability of theseshapes readily in market.

References

- Di Luzio, Frank C. (January 1967). "Water Pollution Control: An American Must". Journal (Water Pollution Control Federation). Water Environment Federation. 39 (1): 1–7. JSTOR 25035710
- Wastewater engineering : treatment and reuse (4th ed.). Metcalf & Eddy, Inc., McGraw Hill, USA. 2003. p. 1807. ISBN 0-07-112250-8

- Vigneswaran, S; Sundaravadivel, M. RECYCLE AND REUSE OF DOMESTIC WASTEWATER (PDF). Retrieved 29 July 2016
- Burian, Steven J.; Nix, Stephan J.; Pitt, Robert E.; Durrans, S. Rocky (2000). "Urban Wastewater Management in the United States: Past, Present, and Future" (PDF). Journal of Urban Technology. London: Routledge. 7 (3). doi:10.1080/713684134
- G. Allen Burton, Jr., Robert Pitt (2001). Stormwater Effects Handbook: A Toolbox for Watershed Managers, Scientists, and Engineers. New York: CRC/Lewis Publishers. ISBN 0-87371-924-7.
- Wachman, Richard (2007-12-09). "Water becomes the new oil as world runs dry". The Guardian. London. Retrieved 2015-09-23.
- Angelakis, Andreas N.; Snyder, Shane A. (9 September 2015). "Wastewater Treatment and Reuse: Past, Present, and Future". Water. 7 (9): 4887–4895. doi:10.3390/w7094887
- Kennish, Michael J. (1992). Ecology of Estuaries: Anthropogenic Effects. Marine Science Series. Boca Raton, FL: CRC Press. pp. 415–17. ISBN 978-0-8493-8041-9
- "Maya plumbing: First pressurized water feature found in New World". Penn State. May 5, 2010. Retrieved 26 March 2014
- Tzanakakis, V. E.; Paranychianaki, N. V.; Angelakis, A. N. (1 March 2007). "Soil as a wastewater treatment system: historical development". Water Science and Technology: Water Supply. 7 (1): 67–75. ISSN 1606-9749. doi:10.2166/ws.2007.008
- Laws, Edward A. (2000). Aquatic Pollution: An Introductory Text. New York: John Wiley and Sons. p. 430. ISBN 978-0-471-34875-7
- Khouri, N; Kalbermatten, J. M.; Bartone, C. R. "Reuse of wastewater in agriculture: A guide for planners" (PDF). Retrieved 29 July 2016
- Moss, Brian (2008). "Water Pollution by Agriculture" (PDF). Phil. Trans. Royal Society B. 363: 659–666. doi:10.1098/rstb.2007.2176
- Ghneim, Azmi (2010). Wastewater reuse und management in the Middle East and North Africa a case study of Jordan ([Online-Ausg.]. ed.). Berlin: Univ.-Verl. der TU. ISBN 3798322686
- Shuval, H. "Wastewater recycling and reuse as a water source for Mediterranean countries: Hygienic and technological aspects". www.oieau.fr. Retrieved 29 July 2016
- Zaikab, Gwyneth Dickey (2011-03-28). "Marine microbes digest plastic". Nature. Macmillan. ISSN 0028-0836. doi:10.1038/news.2011.191
- Evolution of sanitation and wastewater technologies through the centuries. [S.l.]: IWA (International Water Association). 2014. ISBN 9781780404844
- "Ancient Inventions Showers". inventions.org. inventions.org. Archived from the original on May 9, 2008. Retrieved 26 March 2014

2

Water Pollutants: Sources and Effects

Water pollutants can be categorized into seven types. They are inorganic pollutants (salts and metals), organic pollutants, radioactive pollutants, nutrients and agriculture runoff, pathogens, suspended solids and sediments (organic and inorganic) and thermal pollution. Sources of pollution is divided into point source, nonpoint source and groundwater pollution. The chapter strategically encompasses and incorporates the major components and key concepts of wastewater management, providing a complete understanding.

United States Regulation of Point Source Water Pollution

Point source water pollution comes from discrete conveyances and alters the chemical, biological, and physical characteristics of water. In the United States, it is largely regulated by the Clean Water Act (CWA). Among other things, the Act requires dischargers to obtain a National Pollutant Discharge Elimination System (NPDES) permit to legally discharge pollutants into a water body. However, point source pollution remains an issue in some water bodies, due to some limitations of the Act. Consequently, other regulatory approaches have emerged, such as water quality trading and voluntary community-level efforts.

Introduction

Definition

Water pollution is the contamination of natural water bodies by chemical, physical, radioactive or pathogenic microbial substances. Point sources of water pollution are described by the CWA as "any discernible, confined, and discrete conveyance from which pollutants are or may be discharged." These include pipes or man-made ditches from stationary locations such as sewage treatment plants, factories, industrial wastewater treatment facilities, septic systems, ships, and other sources that are clearly discharging pollutants into water sources.

Relevant Science

The input of pollutants into a water body may impact the water's ability to deliver ecological, recreational, educational, and economic services. While the impacts of water pollution vary considerably based on a variety of site-specific factors, they may be either direct or indirect. Pollutants that are directly toxic pose a threat to organisms that may come into contact with contaminated water. These include persistent organic pollutants used as pesticides and toxic byproducts of industrial activity (such as cyanide). Other pollutants may indirectly impact ecosystem services by causing a change in water conditions that allows for a harmful activity to take place. This includes sediment (loose soil) inputs that decrease the amount of light that can penetrate through the water, reducing plant growth and diminishing oxygen availability for other aquatic organisms.

There are a variety of water quality parameters that may be affected by point source water pollution. They include: dissolved oxygen and biochemical oxygen demand (BOD), temperature, pH, turbidity, phosphorus, nitrates, total suspended solids, conductivity, alkalinity, and fecal coliform. Given that much of the point source water pollution in the United States comes from municipal wastewater treatment plants, BOD is perhaps the most widely used metric to assess water quality.

Water quality is also closely linked with water quantity issues. Water shortages from natural and anthropogenic activity reduce the dilutive properties of water and may concentrate water pollution. Oppositely, during flooding events, water pollution may spread to previously uncontaminated waters through surface overflow or the failure of man-made barriers.

Nature of the Problem/Context

Cuyahoga River Fire

Before the CWA was enacted, companies indiscriminately discharged their effluents into water bodies. One such water body was the Cuyahoga River located in north-east Ohio. The river was thrust into the national limelight in 1969 when it caught fire, although the river had been plagued by fires since 1936. Pollution of the river had become prevalent in the early 1800s as contaminants from municipal and industrial discharges, bank erosion, commercial/residential development, atmospheric deposition, hazardous waste disposal sites, urban storm water runoff, combined sewer overflows (CSOs) and wastewater treatment plant bypasses were discharged into the river.*Time* magazine described the Cuyahoga as the river that “oozes rather than flows” and in which a person “does not drown but decays.” The 1969 fire drew significant public attention to the state of the nation’s waterways and is sometimes credited for the creation of the Federal Water Pollution Control Act (1972), commonly called the Clean Water Act, the Oil Pollution Act of 1990, and the establishment of federal and state agencies such as the Environmental Protection Agency (EPA).

Regulatory Framework

History of Regulation

Historically, regulation of point source water pollution in the United States included health- and use-based standards to protect environmental and economic interests. The Rivers and Harbors Act of 1899 contained provisions that made discharging refuse matter into navigable waters of the United States illegal without a permit issued by the U.S. Army Corps of Engineers.

In 1948 Congress passed the Federal Water Pollution Control Act (FWPCA). Although it was amended several times, the original FWPCA granted the Surgeon General of the Public Health Service the authority to develop programs to combat pollution that was harming surface and underground water sources. The FWPCA also authorized cooperation between federal and state agencies to construct waste treatment plants.

The FWPCA amendments of 1966, which came to be known as the *Clean Water Restoration Act* established a study to determine the effects of pollution on wildlife, recreation, and water supplies. The Act also set forth guidelines for abatement of water that may flow into international territory and prohibited the dumping of oil into navigable waters of the United States.

Further amendments to FWPCA in 1970 were dubbed the *Water Quality Improvement Act*. They required the development of certain water quality standards and expanded federal authority in upholding the standards. The most substantial amendments to the FWPCA occurred in 1972 and became known as the *Clean Water Act*.

Clean Water Act

Point source water pollution is largely regulated through the Clean Water Act, which gives the EPA the authority to set limits on the acceptable amount of pollutants that can be discharged into waters of the United States. The Act broadly defines a pollutant as any type of industrial, municipal, and agricultural waste discharged into water, such as: dredged soil, solid waste, incinerator residue, sewage, garbage, sewage sludge, munitions, chemical wastes, biological materials, radioactive materials, heat, wrecked or discarded equipment, rock, sand, cellar dirt and industrial, municipal, and agricultural waste. Point source water pollution is discharged into waters through both direct and indirect methods.

Indirect Point Source Water Pollution

An indirect discharger is one that sends its wastewater into a city sewer system, which carries it to the municipal sewage treatment plant or publicly owned treatment works (POTW). At the POTW, harmful pollutants in domestic sewage, called conventional pollutants, are removed from the sewage and then discharged into a surface water. POTWs are not designed to treat toxic or nonconventional pollutants in industrial wastewater, although they may incidentally remove some pollutants.

National Pretreatment Program

Indirect dischargers are covered by the National Pretreatment Program, which enforces three types of discharge standards:

- prohibited discharge standards – protect against pass-through and interference
- categorical standards – national, uniform, technology-based standards that limit the discharge of pollutants
- local limits – address the specific needs of a POTW and its receiving waters.

The goal of the pretreatment program is to protect municipal wastewater treatment plants from damage that may occur when hazardous, toxic, or other wastes are discharged into a sewer system and to protect the quality of sludge generated by these plants. Discharges to a POTW are regulated primarily by the POTW itself, rather than the state/tribe or EPA.

Direct Point Source Water Pollution

Direct discharges are pollutants that are discharged directly into the water. To legally discharge pollutants directly into a waterbody, a National Pollution Discharge Elimination System (NPDES) permit must be obtained.

Deer Island Waste Water Treatment Plant, serving the Boston, Massachusetts area

NPDES Permit Program

The NPDES permit program sets limits on the amount of pollutants that can be discharged into a waterbody. Technology based effluent limits establish a minimum level of pollution controls for all point source discharges. If technology based limits are not sufficient, water quality based effluent limits are developed.

Permits

Individual states are authorized by the EPA to issue permits when they have demonstrated that their program is at least as stringent as the EPA's program. States perform the day-to-day issuance of permits and oversight of the program while the EPA provides review and guidance to the states. All NPDES permits must contain a "specific, numeric, measurable set of limits on the amount of various pollutants that can appear in the wastewater discharged by the facility into the nation's waters" as well as guidelines on how often monitoring should be performed and what "sampling and analytic techniques should be used."

Types of Permits

- Individual – A unique permit is issued for each discharger.
- General – A single permit that covers a large number of similar dischargers in a specific geographic area.

Permitting Process

The authorized permit issuing body receives and reviews the permit application. The technology-based and water quality-based effluent limits are developed and then compared to determine which is the more stringent, which is then used as the effluent limit for the permit. Monitoring requirements, special conditions, and standard conditions for each pollutant are developed and the permit is then issued and its requirements are implemented.

- Technology-based requirements: A minimum level of treatment based on available treatment technologies is required for discharged pollutants, however, a discharger may use any available treatment technologies to meet the limits. The effluent limits are derived from different standards for different discharges:
  - o Municipal discharges (POTW): National secondary treatment standards define limits of biological treatment standards based on biochemical oxygen demand (BOD), total suspended solids (TSS), and pH balance.
  - o Non-municipal discharges: Limits for non-municipal (e.g. industrial) discharges are based on national discharge standards for industrial facilities within a certain category. These limits are achieved for using pollution control and prevention technologies designed for different types of dischargers. For existing dischargers, this level of treatment is equivalent to "Best Available Technology Economically Achievable" (BAT) and for new discharges, the treatment level is "New Source Performance Standards" (NSPS).
- Water quality-based requirements: Should technology-based standards not be stringent enough, Water Quality Based Effluent Limits (WQBEL) are developed to ensure that water quality standards are attained. WQBELs are based on ambient water quality standards.

Permit Components

All NPDES permits must contain the following five components:

- Cover page – indicates authorization for discharging and its locations
- Effluent limits – limits used to control discharges through technology-based or water quality-based standards
- Monitoring and reporting requirements – used to determine permit compliance
- Special conditions – can be used to supplement effluent limits
- Standard conditions – pre-established conditions that apply to all NPDES permits

Permit Violation

A permitee can be in violation of their permit when they discharge pollutants at a level higher than what is specified on their permit or discharge without a permit. They can also be in violation if they fail to comply with the monitoring and enforcement portion of the permit.

Enforcement

Since the NPDES permit program is a self-monitoring system where permitees are required to carry out detailed monitoring requirements, the EPA promotes "compliance assistance" as an enforcement technique, which "helps permittees come into, and remain, in compliance with their permit, rather than going immediately to enforcement actions." The EPA and state NPDES agencies have can perform periodic inspections and the EPA gives individual states the authority to enforce NP-

DES permits although the EPA has the right to carry out enforcement should a state not do so. Enforcement actions for violations include: injunctions, fines for typical violations, imprisonment for criminal violations, or supplemental environmental projects (SEP). Citizens may also bring suits against violators but they must first provide the EPA and state NPDES permit agency with the opportunity to take action.

Stormwater Management Permits

A silt fence, a type of sediment control, installed on a construction site

To address the nationwide problem of stormwater pollution, Congress broadened the CWA definition of "point source" in 1987 to include industrial stormwater discharges and municipal separate storm sewer systems ("MS4"). These facilities were required to obtain NPDES permits. This 1987 expansion was promulgated in two phases. The Phase I regulation, promulgated in 1990, required that all municipalities of 100,000 persons or more, industrial dischargers, and construction sites of 5 acres (20,000 m²) or more have NPDES permits for their stormwater discharges. Phase I permits were issued in much of the U.S. in 1991. The Phase II rule required that all municipalities, construction sites of 1 acre (4,000 m²) or more, and other large property owners (such as school districts) have NPDES permits for their stormwater discharges. EPA published the Phase II regulation in 1999.

Most construction sites are covered by general permits. Other industrial sites that only discharge stormwater are typically covered by general permits.Industrial stormwater dischargers that are otherwise required to have individual permits (due to their process wastewater and/or cooling water discharges), typically have the stormwater management requirements added to their individual permits.

In addition to implementing the NPDES requirements, many states and local governments have enacted their own stormwater management laws and ordinances, and some have published stormwater treatment design manuals. Some of these state and local requirements have expanded coverage beyond the federal requirements. For example, the State of Maryland requires erosion and sediment controls on construction sites of 5,000 sqft (460 m²) or more. It is not uncommon for state agencies to revise their requirements and impose them upon counties and cities; daily fines ranging as high as $25,000 can be imposed for failure to modify their local stormwater permitting for construction sites, for instance.

Others

EPA Water Quality Trading Policy

The Clean Water Act has made great strides in reducing point source water pollution, but this effect is overshadowed by the fact that nonpoint source pollution, which is not subject to regulation under the Act, has correspondingly increased. One of the solutions to address this imbalance is point/nonpoint source trading of pollutants. In January 2003, the EPA Water Quality Trading Policy was issued. At this time, many waters in the United States did not support their designated uses. Specifically, 40 percent of rivers, 45 percent of streams, and 50 percent of lakes that had been surveyed were unfit. Consequently, when The Water Quality Trading Policy was issued it acknowledged that "the progress made toward restoring and maintaining the chemical, physical, and biological integrity of the nation's waters under the 1972 Clean Water Act and its National Pollutant Discharge Elimination System (NPDES) permits has been incomplete."

The purpose of the policy is to "encourage voluntary trading programs that facilitate the implementation of TMDLs, reduce the costs of compliance with CWA regulations, establish incentives for voluntary reductions, and promote watershed-based initiatives (3)." The policy supports the trading of nutrients such as nitrogen and phosphorus and sediment load reductions, but in order for it to be extended to other contaminants, more scrutiny is required. All water quality trading programs are subject to the requirements of the Clean Water Act.

The Trading Policy outlines basic ground rules for trading by specifying viable pollutants, how to set baselines, and detailing the components of credible trading programs. It also stipulates that trades must occur within the same watershed. Water quality trading programs are subject to the stipulations of the Clean Water Act.

Other Laws that may Affect some NPDES Permits

- Endangered Species Act: Federal agencies must consult with the U.S. Fish and Wildlife Service to ensure that discharges from a project (e.g., a new or expanded industrial facility) will not endanger a threatened species or their habitat.
- National Environmental Policy Act (NEPA): Only discharges that are subject to New Source Performance Standards (or new sources otherwise defined in the NPDES regulations) are subject to NEPA review prior to being issued a permit.
- National Historic Preservation Act: Prior to issuing a permit, EPA Regional Administrators must adopt measures that mitigate adverse effects on properties in the National Register of Historic Places.
- Coastal Zone Management Act: Permits will not be issued unless the permitees certifies that proposed activities, which would affect land or water use in coastal zones, comply with the Coastal Zone Management Act.
- Wild and Scenic Rivers Act: Prohibits issuance of permit for water resources projects that will have a direct, adverse effect on the values for which a national wild and scenic river was established.

- Fish and Wildlife Coordination Act: Jurisdiction over wildlife resources must be established prior to permit issuance so that resources can be conserved.
- *Magnuson-Stevens Act:* The "Essential Fish Habitat Provisions" in the law require EPA to consult with the National Marine Fisheries Service for any EPA-issued permits which may adversely affect essential fish habitat.

Problems/Issues/Concerns

Cost Issues for Monitoring

Monitoring of water bodies is the responsibility of authorized states, not the EPA. In 1997, EPA estimated that private and public point source control costs were $14 billion and $34 billion, respectively. The EPA has acknowledged that states have not adequately funded their monitoring programs, which has led to some uncertainty regarding the quality of most surface waters.

Enforcement

Self-monitoring and Self-reporting

In many cases, the enforcement mechanisms of the Clean Water Act have created tension between regulators, regulated parties, and local citizens. Most NPDES permits require dischargers to monitor and report the contents of their discharges to the appropriate authorities. This requirement is potentially self-incriminating, forcing industries to provide information that may subject them to penalties and legal constraints. As a result, some dischargers go to great lengths to avoid penalties, including falsifying discharge monitoring reports and tampering with monitoring equipment. In *United States v. Hopkins*, the court ruled on a case where the vice president for manufacturing at Spirol International Corporation was charged with three criminal violations for falsifying water samples sent to state regulatory agencies.Spirol diluted his samples, which contained high levels of zinc, with tap water on numerous occasions and frequently ordered his subordinates to reduce the zinc concentration in the water by running it through a coffee filter.

Tensions between State and Federal Government

Like other environmental laws, the Clean Water Act delegates many enforcement responsibilities to state agencies. While the burden of enforcement may be transferred to the states, federal agencies reserve the right to approve or reject state plans for dealing with water pollution. This relationship reduces the regulatory burden on federal agencies, but can lead to confusion and tension between the two regulators.

Many of these tensions arise with regards to the commerce clause of the constitution. Until recently, the commerce clause has given the federal government considerable authority in regulating states' decisions about water use. In 2000, the United States Supreme Court ruled on *Solid Waste Agency of Northern Cook County v. US Army Corps of Engineers*. This ruling struck down the Corps' ability to prevent the construction of a disposal site for non-hazardous waste in Illinois based on power derived from the commerce clause. The Corps cited the Migratory Bird Rule when they initially denied the section 404 permit under the Clean Water Act. The migratory bird rule was meant to protect habitats used by migratory birds, which included the abandoned mining site that SWANCC had proposed to construct the waste disposal site. Chief Justice Rehnquist wrote: "Congress passes the CWA for the state purpose of 'restoring and maintaining the chemical, physical, and biological integrity of the Nation's waters.' In doing so,, Congress chose to recognize, preserve, and protect the primary responsibilities and rights of States to prevent, reduce, and eliminate pollution, to plan the development and use... of land and water resources...". In reversing the Corps' decision to issue a permit, the court reversed a trend and placed a check on federal power over state land use and water rights. Tensions between federal and state agencies concerning interstate commerce and point source water pollution continue, and are a reality of the Clean Water Act.

Ambiguity of the CWA

Coeur Alaska v. Southeast Alaska Conservation Council

In 2009, the Supreme Court ruled on *Coeur Alaska, Inc. v. Southeast Alaska Conservation Council.* The case concerned the re-opening of a gold mine outside Juneau, Alaska that had been out of operation since 1928. Coeur Alaska planned to utilize froth flotation in order to extract gold, creating 4.5 million tons of tailings over the course of its lifetime. The mining company opted to dispose of the tailings in nearby Lower Slate Lake, requiring a permit to comply with the Clean Water Act. The tailings would fundamentally change the physical and chemical characteristics of the lake, raising the lake bed by 50 feet and expanding the area from 23 to 60 acres. Coeur Alaska proposed to temporarily re-route nearby streams around Lower Slate Lake until they could purify the water and re-introduce the natural flow patterns.

Tailings from froth flotation contain high concentrations of heavy metals, including aluminum, which have toxic effects of aquatic organisms. As a result, the disposal of these tailings into Lower Slate Lake is eligible for a section 402 permit for discharge of a pollutant from the EPA (NPDES permit). The nature of the tailings also justifies their categorization as a fill material, or a "material [that] has the effect of... changing the bottom elevation" of a water body. Consequently, Coeur Alaska was also eligible for a "Dredge and Fill" permit from the Army Corps of Engineers under CWA section 404. The company applied for this latter permit and received authorization from the Corps to dump the tailings into Lower Slate Lake. The Southeast Alaska Conservation Council contended that disposal of the tailings is explicitly banned by section 306(e) of the Clean Water Act, and would therefore make Coeur Alaska ineligible for a NPDES permit.

The Court ruled in favor of Coeur Alaska, explaining that if the Army Corps of Engineers has authority to issue a permit under section 404, the EPA does not have authority to issue a section 402 permit. They asserted that the law is ambiguous as to whether section 306 applies to fill materials and found no erroneous or unreasonable behavior by the Corps. As a result, although the tailings would explicitly violate the Clean Water Act under section 402, the Corps may issue a dredge and fill permit.

This decision has not resonated well with environmental groups, who are worried that the decision may allow companies to discharge massive amounts of hazardous pollutants by avoiding the NPDES permitting procedure. Of particular concern is the mountaintop mining industry, which has the capacity to fundamentally alter aquatic ecosystems by filling in water bodies with sediment and mining debris. This tension between various sections within the Clean Water Act is sure to receive considerable attention in future years.

Other Emerging Regulatory Approaches

Water Quality Trading

Water quality trading (WQT) is a market-based approach, implemented on a watershed-scale, used to improve or maintain water quality. It involves the voluntary exchange of pollution reduction credits from sources with low costs of pollution control to those with high costs of pollution control. WQT programs are still subject to the requirements of the Clean Water Act, but they can be used to reduce the overall cost of compliance. Usually, permitted point sources of water pollution, such as wastewater treatment plants, have high discharge treatment costs, whereas nonpoint sources of water pollution, such as agriculture, have low costs of pollution reduction. Therefore, it is generally assumed that most trades would take place between point sources and nonpoint sources. However, point source-point source trades could also occur as well as pretreatment trades and intra-plant trades.

Background

Most of the water quality trading markets currently in operation are focused on the trading of nutrients such as phosphorus and nitrogen. However, increasing interest has been shown in trading programs for sediment runoff, biological oxygen demand, and temperature. WQT programs can be used to preserve good water quality in unimpaired waters by counterbalancing new or increased pollutant discharges. In impaired waters, a WQT program can be used to improve water quality by reducing pollutant discharges in order to meet a specified water quality standard or total maximum daily load (TMDL).

TMDLs apply to both point sources and nonpoint sources and they represent the primary impetus for WQT programs. Point sources of pollutants that require NPDES permits often have strict discharge limits based on a TMDL. WQT can allow these sources to obtain lower costs of compliance, while still achieving the overall desired pollution reduction. Several factors influence whether or not a TMDL-based water quality trading program will be successful. First, the market must be appropriately structured within the regulatory framework of the Clean Water Act. Second, the pollutant must be well-suited for trading. Third, implementation of a WQT market requires public input and voluntary participation. Finally, there must be adequate differences in pollution control costs and available opportunities for reduction.

Credits and Trade Ratios

In a WQT market, a unit of pollutant reduction is called a credit. A point source can generate credits by reducing its discharge below its most stringent effluent limitation and a nonpoint source can generate credits "by installing best management practices (BMPs) beyond its baseline". Before being able to purchase credits, source must first meet its technology-based effluent limit (TBEL). The credits can then be used to meet water quality-based effluent limits (WQBEL).

In order to ensure that trades are effective and do not result in more pollution than would occur in their absence, trade ratios are used. Trade ratios can have several components including:

- Location: Source location relevant to the downstream area of concern can be an important factor.

- Delivery: The distance between sources can play a role in determining whether permit requirements are met at the outfall.
- Uncertainty: Nonpoint source reductions can be difficult to quantify.
- Equivalency: Sources may be discharging different forms of the same pollutant.
- Retirement: Credits may be retired to achieve further water quality improvement.

Permitted point sources can trade with other point sources or nonpoint sources. Trades can occur directly, or be brokered by third parties. However, when dealing with nonpoint source reductions, a level of uncertainty does exist. In order to address this, monitoring should be conducted. Modeling can also be used as a supplement to monitoring. Uncertainty can also be mitigated by field testing BMPs and using conservative assumptions for BMP efficacy.

Benefits

There are many economic, environmental, and social benefits that can be gained by establishing a WQT market within a watershed. Economically, since WQT is a market-based policy instrument, substantial savings can be generated while still achieving a mandated water quality goal. According to The National Cost to Implement Total Maximum Daily Loads (TMDLs) Draft report, flexible approaches to improving water quality, such as WQT markets, could save $900 million per year when compared with the least flexible approach (3). In 2008, WQT programs were worth $11 million, but have the potential for rapid growth. Other economic benefits of WQT include a reduction in the overall costs of compliance, the ability for dischargers to take advantages of economies of scale and differences in treatment efficiencies, and the ability to maintain growth without further harming the environment.

The environmental benefits of WQT programs are also numerous. First, habitats and ecosystems are protected and/or improved. Second, water quality objectives are able to be achieved in a timely manner. Third, there is incentive for innovation and creation of pollution prevention technologies. Finally, nonpoint sources are included in solving water quality problems. Social benefits include dialog among watershed stakeholders and incentives for all dischargers to reduce their pollution.

Factors Influencing Success

The success of a WQT market is determined by several factors including the pollutant of interest, physical characteristics of the affected watershed, control costs, trading mechanisms, and stakeholder participation and willingness. It is also important that the desired level of pollution reduction is not so great that all sources must reduce the maximum amount possible because this would eliminate surplus reductions to be used for credits.

Obstacles to Implementation

The biggest obstacle to the widespread adoption of WQT markets is lack of supply and demand. In fact, studies of current water quality trading programs indicate that the typical problems associated with inhibition of water quality trading, such as high transaction costs, poor institutional infrastructure, and uncertain criteria, are being overcome. The main problem is that, under existing regulatory conditions, there are simply not enough willing buyers and sellers. Currently, most nonpoint sources of water pollution are unregulated or, assuming detection occurs, have relatively small consequences for violations. Consequently, nonpoint sources do not have incentive to participate in WQT. For WQT markets to be successful, greater demand is needed for pollution credits. For this to happen, water quality standards need to be clear and enforceable.

Contrast to Emissions Trading

Given the success of the sulfur dioxide emission trading market that was established to combat acid rain, at first glance it seems that this level of success should be easily extended to water quality trading. However, the reason this has not occurred yet comes down to a fundamental difference between water pollution and air pollution, and the process of establishing their respective markets. Establishing an emissions market, in principle, has three steps: (1) set a cap on emissions, (2) allocate portions of the cap to individual firms, and (3) allow each firm to meet its allowance through emission reduction or trade. The difference with water pollution, however, is that the problems that cause local water quality issues differ from those that create regional air pollution problems. Discharges into water are difficult to measure and effects are dependent on a variety of other factors and vary with weather and location.

Dealing with Transboundary Pollution: A Case Study of the U.S./Mexico Border Region

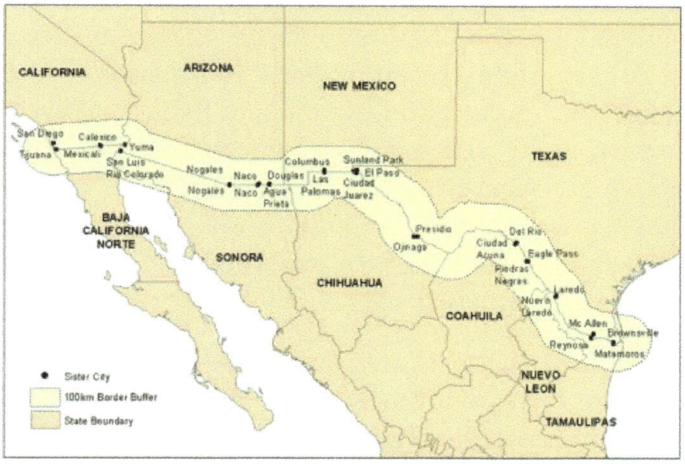

Map of the U.S. – Mexico border region

Overview

The border region (approximately 2,000 miles (3,200 km) long and 62 miles (100 km) wide) is predominantly arid and contains seven watersheds including the Rio Grande which forms part of the border. The watersheds provide numerous benefits for the 14 cities and over nine million people in the region. However, increasing population, the arid climatic conditions of the region, the nature of economic activities along the rivers, increased trade, and uncontrolled emissions into them have placed tremendous pressure on water resources and threatened natural ecosystems. A large proportion of the population lacks access to clean drinking water and sanitation triggering public health concerns.

Policy Issues

Point source water pollution is a source of concern along the US-Mexico border as pollutants from both countries are entering shared waterways due to agricultural runoff, industrial discharge, and untreated sewage. Various policy issues arise in attempting to deal with this and include:

- Some pollution originates from areas beyond the border region as pollution is carried into the region by the waterways. This makes it difficult to regulate as discharges are difficult to apportion and control.
- Pollution is caused by and affects both countries therefore requiring a joint response.
- The socio-economic differences between the two countries have implications for policy implementation and enforcement.
- Various interests are represented with strong influence from environmental and social groups. Multiple levels of government agencies are also involved.

Policy Responses

There have been several attempts to address environmental concerns in the border region in the past by both governments. Significant intervention, however, resulted from the North American Free Trade Agreement (NAFTA) of 1994 between the U.S., Canada, and Mexico which renewed concerns over the environmental quality of the region due to increased trade in the region. The two governments therefore entered into the US-Mexico Border Environment Cooperation Agreement which created a number of institutions and programs. The Border Environmental Cooperation Commission(BECC) and the North American Development Bank (NADB) were created to address border environmental-infrastructure issues and were effected in 2004. Distinct characteristics of these institutions and their approach are that they: are truly bi-national (have members from both countries) at all levels; emphasize a bottom-up approach with enhanced public participation; have a preference to assist disadvantaged communities; avoid regulatory or standard-driven approaches; emphasize economic and environmental sustainability.

The US-Mexico Border Program was also created by the agreement and placed under the management of the EPA (under regions 6 and 9) to correct the oversights of previous institutions and give guidance to cross-border environmental policy. The three institutions work together to identify, develop, finance and implement projects in the communities and certify them as "environmentally sustainable" subsequently funding them through a grant-making process. Communities, public, and private entities (sponsors) are invited to submit water and wastewater infrastructure projects. These projects are required to meet certain criteria to qualify for certification and funding. Among other requirements, they have to address an eligible environmental sector; must have a U.S.-side benefit; and have adequate planning, operations and maintenance, and pretreatment provisions. One specific provision touching on point-source water pollution states that "projects where the discharge is directly or indirectly to U.S. side waters, must target achievement of U.S. norms for ambient water quality in U.S. side waters, although infrastructure development may be phased over time. Any flow reductions that result from implementation of non-discharging alternatives must not threaten U.S. or shared ecosystems". Projects receive significant input from the communities living in the region in determining their sustainability. After certification, the project then receives funding from the NADB. So far, the BECC has certified 160 projects worth approximately three billion dollars since 2004.

The border program has also facilitated direct provision of infrastructure by the federal and state governments such as the construction of wastewater treatment plants, sewer lines, and raw water storage lagoons. One such example is the construction of the Matamoros lift station which is the first phase in eliminating raw sewage discharges into the Rio Grande. Further, the program emphasizes the provision of environmental education and information to communities living in the region. The EPA has established an informational website to provide news and information on the program.

Nonpoint Source Pollution

Muddy river

Nonpoint source (NPS) pollution is a term used to describe pollution resulting from many diffuse sources, in direct contrast to point source pollution which results from a single source. Nonpoint source pollution generally results from land runoff, precipitation, atmospheric deposition, drainage, seepage, or hydrological modification (rainfall or snowmelt) where tracing the pollution back to a single source is difficult.

Non-point source water pollution affects a water body from sources such as polluted runoff from agricultural areas draining into a river, or wind-borne debris blowing out to sea. Non-point source air pollution affects air quality from sources such as smokestacks or car tailpipes. Although these pollutants have originated from a point source, the long-range transport ability and multiple sources of the pollutant make it a non-point source of pollution. Non-point source pollution can be contrasted with point source pollution, where discharges occur to a body of water or into the atmosphere at a single location.

NPS may derive from many different sources with no specific solution may change to rectify the problem, making it difficult to regulate. Non point source water pollution is difficult to control because it comes from the everyday activities of many different people, such as fertilizing a lawn, using a pesticide, or constructing a road or building.

It is the leading cause of water pollution in the United States today, with polluted runoff from agriculture the primary cause.

Other significant sources of runoff include hydrological and habitat modification, and silviculture (forestry).

Contaminated stormwater washed off parking lots, roads and highways, and lawns (often containing fertilizers and pesticides) is called urban runoff. This runoff is often classified as a type of NPS pollution. Some people may also consider it a point source because many times it is channeled into municipal storm drain systems and discharged through pipes to nearby surface waters. However, not all urban runoff flows through storm drain systems before entering water bodies. Some may flow directly into water bodies, especially in developing and suburban areas. Also, unlike other types of point sources, such as industrial discharges, sewage treatment plants and other operations, pollution in urban runoff cannot be attributed to one activity or even group of activities. Therefore, because it is not caused by an easily identified and regulated activity, urban runoff pollution sources are also often treated as true non-point sources as municipalities work to abate them.

Principal Types

Runoff of soil and fertilizer during a rain storm

Sediment

Sediment (loose soil) includes silt (fine particles) and suspended solids (larger particles). Sediment may enter surface waters from eroding stream banks, and from surface runoff due to improper plant cover on urban and rural land. Sediment creates turbidity (cloudiness) in water bodies, reducing the amount of light reaching lower depths, which can inhibit growth of submerged aquatic plants and consequently affect species which are dependent on them, such as fish and shellfish. High turbidity levels also inhibit drinking water purification systems.

Sediment can also be discharged from multiple different sources. Sources include construction sites (although these are point sources, which can be managed with erosion controls and sediment controls), agricultural fields, stream banks, and highly disturbed areas.

Nutrients

Nutrients mainly refers to inorganic matter from runoff, landfills, livestock operations and crop lands. The two primary nutrients of concern are phosphorus and nitrogen.

Phosphorus is a nutrient that occurs in many forms that are bioavailable. It is notoriously over-abundant in human sewage sludge. It is a main ingredient in many fertilizers used for agriculture as well as on residential and commercial properties, and may become a limiting nutrient in freshwater systems and some estuaries. Phosphorus is most often transported to water bodies via soil erosion because many forms of phosphorus tend to be adsorbed on to soil particles. Excess amounts of phosphorus in aquatic systems (particularly freshwater lakes, reservoirs, and ponds) leads to proliferation of microscopic algae called phytoplankton. The increase of organic matter supply due to the excessive growth of the phytoplankton is called eutrophication. A common symptom of eutrophication is algae blooms that can produce unsightly surface scums, shade out beneficial types of plants, and poison the water due to toxins produced by the algae. These toxins are a particular problem in systems used for drinking water because some toxins can cause human illness and removal of the toxins is difficult and expensive. Bacterial decomposition of algal blooms consumes dissolved oxygen in the water, generating hypoxia with detrimental consequences for fish and aquatic invertebrates.

Nitrogen is the other key ingredient in fertilizers, and it generally becomes a pollutant in saltwater or brackish estuarine systems where nitrogen is a limiting nutrient. Similar to phosphorus in fresh-waters, excess amounts of bioavailable nitrogen in marine systems lead to eutrophication and algae blooms. Hypoxia is an increasingly common result of eutrophication in marine systems and can impact large areas of estuaries, bays, and near shore coastal waters. Each summer, hypoxic conditions form in bottom waters where the Mississippi River enters the Gulf of Mexico. During recent summers, the aereal extent of this "dead zone" is comparable to the area of New Jersey and has major detrimental consequences for fisheries in the region.

Nitrogen is most often transported by water as nitrate (NO_3). The nitrogen is usually added to a watershed as organic-N or ammonia (NH_3), so nitrogen stays attached to the soil until oxidation converts it into nitrate. Since the nitrate is generally already incorporated into the soil, the water traveling through the soil (i.e., interflow and tile drainage) is the most likely to transport it, rather than surface runoff.

Toxic Contaminants and Chemicals

Compounds including heavy metals like lead, mercury, zinc, and cadmium, organics like polychlorinated biphenyls (PCBs) and polycyclic aromatic hydrocarbons (PAHs), fire retardants, and other substances are resistant to breakdown. These contaminants can come from a variety of sources including human sewage sludge, mining operations, vehicle emissions, fossil fuel combustion, urban runoff, industrial operations and landfills.

Toxic chemicals mainly include organic compounds and inorganic compounds. These compounds include pesticides like DDT, acids, and salts that have severe effects to the ecosystem and water-bodies. These compounds can threaten the health of both humans and aquatic species while being resistant to environmental breakdown, thus allowing them to persist in the environment.

These toxic chemicals could come from croplands, nurseries, orchards, building sites, gardens, lawns and landfills.

Acids and salts mainly are inorganic pollutants from irrigated lands, mining operations, urban runoff, industrial sites and landfills.

Pathogens

Pathogens are bacteria and viruses that can be found in water and cause diseases in humans. Typically, pathogens cause disease when they are present in public drinking water supplies. Pathogens found in contaminated runoff may include:

- *Cryptosporidium parvum*
- *Giardia lamblia*
- *Salmonella*
- *Novovirus* and other viruses
- Parasitic worms (helminths).

Coliform bacteria and fecal matter may also be detected in runoff. These bacteria are a commonly used indicator of water pollution, but not an actual cause of disease.

Pathogens may contaminate runoff due to poorly managed livestock operations, faulty septic systems, improper handling of pet waste, the over application of human sewage sludge, contaminated storm sewers, and sanitary sewer overflows.

Principal Sources

Urban and Suburban Areas

Urban and suburban areas are a main sources of nonpoint source pollution due to the amount of runoff that is produced due to the large amount of paved surfaces. Paved surfaces, such as asphalt and concrete are impervious to water penetrating them. Any water that is on contact with these surfaces will run off and be absorbed by the surrounding environment. These surfaces make it easier for stormwater to carry pollutants into the surrounding soil.

Construction sites tend to have disturbed soil that is easily eroded by precipitation like rain, snow, and hail. Additionally, discarded debris on the site can be carried away by runoff waters and enter the aquatic environment.

Typically, in suburban areas, chemicals are used for lawn care. These chemicals can end up in runoff and enter the surrounding environment via storm drains in the city. Since the water in storm drains is not treated before flowing into surrounding water bodies, the chemicals enter the water directly.

Agricultural Operations

Agricultural operations account for a large percentage of all nonpoint source pollution in the

United States. When large tracts of land are plowed to grow crops, it exposes and loosens soil that was once buried. This makes the exposed soil more vulnerable to erosion during rainstorms. It also can increase the amount of fertilizer and pesticides carried into nearby bodies of water.

Atmospheric Inputs

Atmospheric inputs of pollution into the air can come from multiple sources. Typically, industrial facilities, like factories, emit air pollution via a smokestack. Although this is a point source, due to the distributional nature, long-range transport, and multiple sources of the pollution, it is considered a nonpoint source. Additionally, atmospheric pollution can become water pollution, by being washed out of the atmosphere in the form of rain or snow.

Highway Runoff

Highway runoff accounts for a small but widespread percentage of all nonpoint source pollution. Harned (1988) estimated that runoff loads were composed of atmospheric fallout (9%), vehicle deposition (25%) and highway maintenance materials (67%) he also estimated that about 9 percent of these loads were reentrained in the atmosphere.

Forestry and Mining Operations

Forestry and mining operations can have significant inputs to non-point source pollution.

Forestry

Forestry operations reduce the number of trees in a given area, thus reducing the oxygen levels in that area as well. This action, coupled with the heavy machinery rolling over the soil increases the risk of erosion.

Mining

Active mining operations are considered point sources, however runoff from abandoned mining operations contribute to nonpoint source pollution. In strip mining operations, the top of the mountain is removed to expose the desired ore. If this area is not properly reclaimed once the mining has finished, soil erosion can occur. Additionally, there can be chemical reactions with the air and newly exposed rock to create acidic runoff. Water that seeps out of abandoned subsurface mines can also be highly acidic. This can seep into the nearest body of water and change the pH in the aquatic environment.

Marinas and Boating Activities

Chemicals used for boat maintenance, like paint, solvents, and oils find their way into water through runoff. Additionally, spilling fuels or leaking fuels directly into the water from boats contribute to nonpoint source pollution. Nutrient and bacteria levels are increased by poorly maintained sanitary waste receptacles on the boat and pump-out stations.

Control

Contour buffer strips used to retain soil and reduce erosion

Urban and Suburban Areas

To control non-point source pollution, many different approaches can be undertaken in both urban and suburban areas. Buffer strips provide a barrier of grass in between impervious paving material like parking lots and roads, and the closest body of water. This allows the soil to absorb any pollution before it enters the local aquatic system. Retention ponds can be built in drainage areas to create an aquatic buffer between runoff pollution and the aquatic environment. Runoff and storm water drain into the retention pond allowing for the contaminants to settle out and become trapped in the pond. The use of porous pavement allows for rain and storm water to drain into the ground beneath the pavement, reducing the amount of runoff that drains directly into the water body. Restoration methods such as constructing wetlands are also used to slow runoff as well as absorb contamination.

Construction sites typically implement simple measures to reduce pollution and runoff. Firstly, sediment or silt fences are erected around construction sites to reduce the amount of sediment and large material draining into the nearby water body. Secondly, laying grass or straw along the border of construction sites also work to reduce nonpoint source pollution.

In areas served by single-home septic systems, local government regulations can force septic system maintenance to ensure compliance with water quality standards. In Washington (state), a novel approach was developed through a creation of a "shellfish protection district" when either a commercial or recreational shellfish bed is downgraded because of ongoing nonpoint source pollution. The shellfish protection district is a geographic area designated by a county to protect water quality and tideland resources, and provides a mechanism to generate local funds for water quality services to control nonpoint sources of pollution. At least two shellfish protection districts in south Puget Sound have instituted septic system operation and maintenance requirements with program fees tied directly to property taxes.

Agricultural Operations

To control sediment and runoff, farmers may utilize erosion controls to reduce runoff flows and retain soil on their fields. Common techniques include contour plowing, crop mulching, crop rotation, planting perennial crops and installing riparian buffers. *Conservation tillage* is a concept used to reduce runoff while planting a new crop. The farmer leaves some crop reside from the previous planting in the ground to help prevent runoff during the planting process.

Nutrients are typically applied to farmland as commercial fertilizer; animal manure; or spraying of municipal or industrial wastewater (effluent) or sludge. Nutrients may also enter runoff from crop residues, irrigation water, wildlife, and atmospheric deposition. Farmers can develop and implement nutrient management plans to reduce excess application of nutrients.

To minimize pesticide impacts, farmers may use Integrated Pest Management (IPM) techniques (which can include biological pest control) to maintain control over pests, reduce reliance on chemical pesticides, and protect water quality.

Forestry Operations

With a well-planned placement of both logging trails, also called skid trails, can reduce the amount of sediment generated. By planning the trails location as far away from the logging activity as possible as well as contouring the trails with the land, it can reduce the amount of loose sediment in the runoff. Additionally, by replanting trees on the land after logging, it provides a structure for the soil to regain stability as well as replaces the logged environment.

Marinas

Installing shut off valves on fuel pumps at a marina dock can help reduce the amount of spillover into the water. Additionally, pump-out stations that are easily accessible to boaters in a marina can provide a clean place in which to dispose of sanitary waste without dumping it directly into the water. Finally, something as simple as having trash containers around a marina can prevent larger objects entering the water.

Groundwater Pollution

Groundwater pollution example in Lusaka, Zambia where the pit latrine in the background is polluting the shallow well in the foreground with pathogens and nitrate.

Groundwater pollution (also called groundwater contamination) occurs when pollutants are released to the ground and make their way down into groundwater. It can also occur naturally due to the presence of a minor and unwanted constituent, contaminant or impurity in the groundwater, in which case it is more likely referred to as contamination rather than pollution.

The pollutant creates a contaminant plume within an aquifer. Movement of water and dispersion within the aquifer spreads the pollutant over a wider area. Its advancing boundary, often called a plume edge, can intersect with groundwater wells or daylight into surface water such as seeps and spring, making the water supplies unsafe for humans and wildlife. The movement of the plume, called a plume front, may be analyzed through a hydrological transport model or groundwater model. Analysis of groundwater pollution may focus on soil characteristics and site geology, hydrogeology, hydrology, and the nature of the contaminants.

Pollution can occur from on-site sanitation systems, landfills, effluent from wastewater treatment plants, leaking sewers, petrol filling stations or from over application of fertilizers in agriculture. Pollution (or contamination) can also occur from naturally occurring contaminants, such as arsenic or fluoride. Using polluted groundwater causes hazards to public health through poisoning or the spread of disease.

Different mechanisms have influence on the transport of pollutants, e.g. diffusion, adsorption, precipitation, decay, in the groundwater. The interaction of groundwater contamination with surface waters is analyzed by use of hydrology transport models.

Pollutant Types

Contaminants found in groundwater cover a broad range of physical, inorganic chemical, organic chemical, bacteriological, and radioactive parameters. Principally, many of the same pollutants that play a role in surface water pollution may also be found in polluted groundwater, although their respective importance may differ.

Arsenic and Fluoride

Arsenic and fluoride have been recognized by the World Health Organization (WHO) as the most serious inorganic contaminants in drinking-water on a worldwide basis.

The metalloid arsenic can occur naturally in groundwater, as seen most frequently in Asia, including in China, India and Bangladesh. In the Ganges Plain of northern India and Bangladesh severe contamination of groundwater by naturally occurring arsenic affects 25% of water wells in the shallower of two regional aquifers.

Arsenic in groundwater can also be present where there are mining operations or mine waste dumps that will leach arsenic.

Natural fluoride in groundwater is of growing concern as deeper groundwater is being used, “with more than 200 million people at risk of drinking water with elevated concentrations.” Fluoride can especially be released from acidic volcanic rocks and dispersed volcanic ash when water hardness is low. High levels of fluoride in groundwater is a serious problem in the Argentinean Pampas, Chile, Mexico, India, Pakistan, the East African Rift, and some volcanic islands (Tenerife)

In areas that have naturally occurring high levels of fluoride in groundwater which is used for drinking water, both dental and skeletal fluorosis can be prevalent and severe.

Pathogens

Waterborne diseases can be spread via a groundwater well which is contaminated with fecal pathogens from pit latrines

Pathogens contained in feces can lead to groundwater pollution when they are given the opportunity to reach the groundwater, making it unsafe for drinking. Of the four pathogen types that are present in feces (bacteria, viruses, protozoa and helminths or helminth eggs), the first three can be commonly found in polluted groundwater, whereas the relatively large helminth eggs are usually filtered out by the soil matrix.

Groundwater that is contaminated with pathogens can lead to fatal fecal-oral transmission of diseases (e.g. cholera, diarrhoea).

Nitrate

Nitrate is the most common chemical contaminant in the world's groundwater and aquifers. In some low-income countries nitrate levels in groundwater are extremely high, causing significant health problems. It is also stable (it does not degrade) under high oxygen conditions.

Nitrate levels above 10 mg/L (10 ppm) in groundwater can cause "blue baby syndrome" (acquired methemoglobinemia).Drinking water quality standards in the European Union stipulate less than 50 mg/L for nitrate in drinking water.

However, the linkages between nitrates in drinking water and blue baby syndrome have been disputed in other studies. The syndrome outbreaks might be due to other factors than elevated nitrate concentrations in drinking water.

Elevated nitrate levels in groundwater can be caused by on-site sanitation, sewage sludge disposal and agricultural activities. It can therefore have an urban or agricultural origin.

Organic Compounds

Volatile organic compounds (VOCs) are a dangerous contaminant of groundwater. They are generally introduced to the environment through careless industrial practices. Many of these compounds were not known to be harmful until the late 1960s and it was some time before regular testing of groundwater identified these substances in drinking water sources.

Primary VOC pollutants found in groundwater include aromatic hydrocarbons such as BTEX compounds (benzene, toluene, ethylbenzene and xylenes), and chlorinated solvents including tetrachloroethylene (PCE), trichloroethylene (TCE), and vinyl chloride (VC). BTEX are important components of gasoline. PCE and TCE are industrial solvents historically used in dry cleaning processes and as a metal degreaser, respectively.

Other organic pollutants present in groundwater and derived from industrial operations are the polycyclic aromatic hydrocarbons (PAHs). Due to its molecular weight, Naphthalene is the most soluble and mobile PAH found in groundwater, whereas benzo(a)pyrene is the most toxic one. PAHs are generally produced as byproducts by incomplete combustion of organic matter.

Organic pollutants can also be found in groundwater as insecticides and herbicides. As many other synthetic organic compounds, most pesticides have very complex molecular structures. This complexity determines the water solubility, adsorption capacity, and mobility of pesticides in the groundwater system. Thus, some types of pesticides are more mobile than others so they can more easily reach a drinking-water source.

Metals

Several trace metals can occurs naturally in some rocks or enter in the environmental from natural processes such as weathering. However, industrial activities such as mining and metallurgy, solid waste disposal, painting and enamel works can lead to severe groundwater pollution with elevated concentrations of toxic metals including lead, cadmium and chromium.

The migration of metals (and metalloids) in groundwater will be affected by several factors, in particular by chemical reactions which determine the partitioning of contaminants among different phases and species. Thus, the mobility of metals primarily depends on the pH and redox state of groundwater.

Others

Other organic pollutants include a range of organohalides and other chemical compounds, petroleum hydrocarbons, various chemical compounds found in personal hygiene and cosmetic products, drug pollution involving pharmaceutical drugs and their metabolites. Inorganic pollutants might include other nutrients such as ammonia and phosphate, and radionuclides such as uranium (U) or radon (Rn) naturally present in some geological formations. Saltwater intrusion is also an example of natural contamination, but is very often intensified by human activities.

Groundwater pollution is a worldwide issue. A study of the groundwater quality of the principal aquifers of the United States conducted between 1991 and 2004, showed that 23% of domestic wells had contaminants at levels greater than human-health benchmarks. Another study suggested that the major groundwater pollution problems in Africa, considering the order of importance are: (1) nitrate pollution, (2) pathogenic agents, (3) organic pollution, (4) salinization, and (5) acid mine drainage.

Causes

Naturally-occurring (Geogenic)

"Geogenic" refers to naturally occurring as a result from geological processes.

The natural arsenic pollution occurs because aquifer sediments contain organic matter that generates anaerobic conditions in the aquifer. These conditions result in the microbial dissolution of iron oxides in the sediment and, thus, the release of the arsenic, normally strongly bound to iron oxides, into the water. As a consequence, arsenic-rich groundwater is often iron-rich, although secondary processes often obscure the association of dissolved arsenic and dissolved iron. Arsenic is found in groundwater most commonly as the reduced species arsenite and the oxidized species arsenate, being the acute toxicity of arsenite somewhat greater than that of arsenate. Investigations by WHO indicated that 20% of 25,000 boreholes tested in Bangladesh had arsenic concentrations exceeding 50 μg/l.

The occurrence of fluoride is close related to the abundance and solubility of fluoride-containing minerals such as fluorite (CaF2). Considerably high concentrations of fluoride in groundwater are typically caused by a lack of calcium in the aquifer. Health problems associated with dental fluorosis may occur when fluoride concentrations in groundwater exceed 1.5 mg/l, which is the WHO guideline value since 1984.

The Swiss Federal Institute of Aquatic Science and Technology (EAWAG) has recently developed the interactive Groundwater Assessment Platform (GAP), where the geogenic risk of contamination in a given area can be estimated using geological, topographical and other environmental data without having to test samples from every single groundwater resource. This tool also allows the user to produce probability risk mapping for both arsenic and fluoride.

High concentrations of parameters like salinity, iron, manganese, uranium, radon and chromium, in groundwater, may also be of geogenic origin. This contaminants can be important locally but they are not as widespread as arsenic and fluoride.

On-site Sanitation Systems

A traditional housing compound near Herat, Afghanistan, where a shallow water supply well (foreground) is in close proximity to the pit latrine (behind the white greenhouse) leading to contamination of the groundwater.

Groundwater pollution with pathogens and nitrate can also occur from the liquids infiltrating into the ground from on-site sanitation systems such as pit latrines and septic tanks, depending on the population density and the hydrogeological conditions.

Factors controlling the fate and transport of pathogens are quite complex and the interaction among them is not well understood. If the local hydrogeological conditions (which can vary within a space of a few square kilometres) are ignored, simple on-site sanitation infrastructures such as pit latrines can cause significant public health risks via contaminated groundwater.

Liquids leach from the pit and pass the unsaturated soil zone (which is not completely filled with water). Subsequently, these liquids from the pit enter the groundwater where they may lead to groundwater pollution. This is a problem if a nearby water well is used to supply groundwater for drinking water purposes. During the passage in the soil, pathogens can die off or be adsorbed significantly, mostly depending on the travel time between the pit and the well. Most, but not all pathogens die within 50 days of travel through the subsurface.

The degree of pathogen removal strongly varies with soil type, aquifer type, distance and other environmental factors. For example, the unsaturated zone becomes "washed" during extended periods of heavy rain, providing hydraulic pathway for the quick pass of pathogens. It is difficult to estimate the safe distance between a pit latrine or a septic tank and a water source. In any case, such recommendations about the safe distance are mostly ignored by those building pit latrines. In addition, household plots are of a limited size and therefore pit latrines are often built much closer to groundwater wells than what can be regarded as safe. This results in groundwater pollution and household members falling sick when using this groundwater as a source of drinking water.

Sewage (Treated and Untreated)

Groundwater pollution can be caused by untreated waste discharge leading to diseases like skin lesions, bloody diarrhea and dermatitis. This is more common in locations having limited wastewater treatment infrastructure, or where there are systematic failures of the on-site sewage disposal system. Along with pathogens and nutrients, untreated sewage can also have an important load of heavy metals that may seep into the groundwater system.

The treated effluent from sewage treatment plants may also reach the aquifer if the effluent is infiltrated or discharged to local surface water bodies. Therefore, those substances that are not removed in conventional sewage treatment plants may reach the groundwater as well. For example, detected concentrations of pharmaceutical residues in groundwater were in the order of 50 ng/L in several locations in Germany. This is because in conventional sewage treatment plants, micro-pollutants such as hormones, pharmaceutical residues and other micro-pollutants contained in urine and feces are only partially removed and the remainder is discharged into surface water, from where it may also reach the groundwater.

Groundwater pollution can also occur from leaking sewers which has been observed for example in Germany. This can also lead to potential cross-contamination of drinking-water supplies.

Spreading wastewater or sewage sludge in agriculture may also be included as sources of faecal contamination in groundwater.

Fertilizers and Pesticides

Nitrate can also enter the groundwater via excessive use of fertilizers, including manure spreading. This is because only a fraction of the nitrogen-based fertilizers is converted to produce and other plant matter. The remainder accumulates in the soil or lost as run-off. High application rates of nitrogen-containing fertilizers combined with the high water-solubility of nitrate leads to increased runoff into surface water as well as leaching into groundwater, thereby causing groundwater pollution. The excessive use of nitrogen-containing fertilizers (be they synthetic or natural) is particularly damaging, as much of the nitrogen that is not taken up by plants is transformed into nitrate which is easily leached.

The nutrients, especially nitrates, in fertilizers can cause problems for natural habitats and for human health if they are washed off soil into watercourses or leached through soil into groundwater. The heavy use of nitrogenous fertilizers in cropping systems is the largest contributor to anthropogenic nitrogen in groundwater worldwide.

Feedlots/animal corrals can also lead to the potential leaching of nitrogen and metals to groundwater. Over application of animal manure may also result in groundwater pollution with pharmaceutical residues derived from veterinary drugs.

The US Environmental Protection Agency (EPA) and the European Commission are seriously dealing with the nitrate problem related to agricultural development, as a major water supply problem that requires appropriate management and governance.

Runoff of pesticides may leach into groundwater causing human health problems from contaminated water wells. Pesticide concentrations found in groundwater are typically low, and often the regulatory human health-based limits exceeded are also very low. The organophosphorus insecticide monocrotophos (MCP) appears to be one of a few hazardous, persistent, soluble and mobile (it does not bind with minerals in soils) pesticides able to reach a drinking-water source. In general, more pesticide compounds are being detected as groundwater quality monitoring programs have become more extensive; however, much less monitoring has been conducted in developing countries due to the high analysis costs.

Commercial and Industrial Leaks

A wide variety of both inorganic and organic pollutants have been found in aquifers underlying commercial and industrial activities.

Ore mining and metal processing facilities are the primary responsible of the presence of metals in groundwater of anthropogenic origin, including arsenic. The low pH associated with acid mine drainage (AMD) contributes to the solubility of potential toxic metals that can eventually enter the groundwater system.

There is a increasing concern over the groundwater pollution by gasoline leaked from petroleum underground storage tanks (USTs) of gas stations. BTEX compounds are the most common additives of the gasoline. BTEX compounds, including benzene, have densities lower than water (1 g/ml). Similar to the oil spills on the sea, the non-miscible phase, referred to as Light Non-Aqueous Phase Liquid (LNAPL), will "float" upon the water table in the aquifer.

Chlorinated solvents are used in nearly any industrial practice where degreasing removers are required.PCE is a highly utilized solvent in the dry cleaning industry because of its cleaning effectiveness and relatively low cost. It has also been used for metal-degreasing operations. Because it is is highly volatile, it is more frequently found in groundwater than in surface water. TCE has historically been used as a metal cleaning. The military facility Anniston Army Dept (ANAD) in the United States was placed on the US EPASuperfundNational Priorities List (NPL) because of groundwater contamination with as much as 27 million pounds of TCE. Both PCE and TCE may degrade to vinyl chloride (VC), the most toxic chlorinated hydrocarbon.

Many types of solvents may have also been disposed illegally, leaking over time to the groundwater system.

Chlorinated solvents such as PCE and TCE have densities higher than water and the non-miscible phase is referred to as Dense Non-Aqueous Phase Liquids (DNAPL). Once they reach the aquifer, they will "sink" and eventually accumulate on the top of low-permeability layers.

Historically, wood-treating facilities have also release insecticides such as pentachlorophenol (PCP) and creosote into the environment, impacting the groundwater resources. PCP is aa highly soluble and toxic obsolete pesticide recently listed in the Stockholm Convention on Persistent Organic Pollutants. PAHs and other semi-VOCs are the common contaminants associated with creosote.

Although non-miscible, both LNAPLs and DNAPLs still have the potential to slowly dissolve into the aqueous (miscible) phase to create a plume and thus become a long-term source of contamination. DNAPLs (chlorinated solvents, heavy PAHs, creosote, PCBs) tend to be difficult to manage as they can reside very deep in the groundwater system.

Hydraulic Fracturing

The recent growth of Hydraulic Fracturing ("Fracking") wells in the United States has raised concerns regarding its potential risks of contaminating groundwater resources. The Environmental Protection Agency (EPA), along with many other researchers, has been delegated to study the relationship between hydraulic fracturing and drinking water resources. While it is possible to perform hydraulic fracturing without having a relevant impact on groundwater resources if stringent controls und quality management measures are in place, there are a number of cases where groundwater pollution due to improper handling or technical failures was observed.

While the EPA has not found significant evidence of a widespread, systematic impact on drinking water by hydraulic fracturing, this may be due to insufficient systematic pre- and post- hydraulic fracturing data on drinking water quality, and the presence of other agents of contamination that preclude the link between shale oil/gas extraction and its impact.

Despite the EPA's lack of profound widespread evidence, other researchers have made significant observations of rising groundwater contamination in close proximity to major shale oil/gas drilling sites located in Marcellus*Ellsworth, William (2013). "Injection-Induced Earthquakes"*. Within one kilometer of these specific sites, a subset of shallow drinking water consistently showed higher concentration levels of methane, ethane, and propane concentrations than normal. An evaluation of higher Helium and other noble gas concentration along with the rise of hydrocarbon levels supports the distinction between hydraulic fracturing fugitive gas and naturally occurring "background" hydrocarbon content. This contamination is speculated to be the result of leaky, failing, or improperly installed gas well casings.

Furthermore, it is theorized that contamination could also result from the capillary migration of deep residual hyper-saline water and hydraulic fracturing fluid, slowly flowing through faults and fractures until finally making contact with groundwater resources; however, many researchers argue that the permeability of rocks overlying shale formations are too low to allow this to ever happen sufficiently. To ultimately prove this theory, there would have to be traces of toxic trihalomethanes (THM) since they are often associated with the presence of stray gas contamination, and typically co-occur with high halogen concentrations in hyper-saline waters. Besides, highly saline waters are a common natural feature in deep groundwater systems.

While conclusions regarding groundwater pollution as the result to hydraulic fracturing fluid flow is restricted in both space and time, researchers have hypothesized that the potential for systematic stray gas contamination depends mainly on the integrity of the shale oil/gas well structure, along with its relative geological location to local fracture systems that could potentially provide flow paths for fugitive gas migration.

Though widespread, systematic contamination by hydraulic fracturing has been heavily disputed, one major source of contamination that has the most consensus among researchers of being the most problematic is site-specific accidental spillage of hydraulic fracturing fluid and produced water. So far, a significant majority of groundwater contamination events are derived from surface-level anthropogenic routes rather than the subsurface flow from underlying shale formations. Examples of such events include: a fracking fluid spillage in Acorn Fork Creek, Kentucky that caused a widespread death among aquatic species in 2007; a 420,000 gallon spillage of hyper-saline produced water that turned a once very-fertile farmland in New Mexico into a dead-zone in 2010; and a 42,000 gallon fracking fluid spillage in Arlington, Texas that necessitated an evacuation of over a 100 homes in 2015. While the damage can be obvious, and much more effort is being done to prevent these accidents from occurring so frequently, the lack of data from fracking oil spills continue to leave researchers in the dark. In many of these events, the data acquired from the leakage or spillage is often very vague, and thus would lead researchers to lacking conclusions.

Researchers from the Federal Institute for Geosciences and Natural Resources (BGR) conducted a modelling study for a deep shale-gas formation in the North German Basin. They concluded that the probability is small that the rise of fracking fluids through the geological underground to the surface will impact shallow groundwater.

Landfill Leachate

Leachate from sanitary landfills can lead to groundwater pollution.

Love Canal was one of the most widely known examples of groundwater pollution. In 1978, residents of the Love Canal neighborhood in upstate New York noticed high rates of cancer and an alarming number of birth defects. This was eventually traced to organic solvents and dioxins from an industrial landfill that the neighborhood had been built over and around, which had then infiltrated into the water supply and evaporated in basements to further contaminate the air. Eight hundred families were reimbursed for their homes and moved, after extensive legal battles and media coverage.

Other

Further causes of groundwater pollution are chemical spills from commercial or industrial operations, chemical spills occurring during transport (e.g. spillage of diesel fuels), illegal waste dumping, infiltration from urban runoff or mining operations, road salts, de-icing chemicals from airports and even atmospheric contaminants since groundwater is part of the hydrologic cycle.

The burial of corpses and their subsequent degradation may also pose a risk of pollution to groundwater.

Mechanisms

The passage of water through the subsurface can provide a reliable natural barrier to contamination but it only works under favorable conditions.

The stratigraphy of the area plays an important role in the transport of pollutants. An area can have layers of sandy soil, fractured bedrock, clay, or hardpan. Areas of karst topography on limestone bedrock are sometimes vulnerable to surface pollution from groundwater. Earthquake faults can also be entry routes for downward contaminant entry. Water table conditions are of great importance for drinking water supplies, agricultural irrigation, waste disposal (including nuclear waste), wildlife habitat, and other ecological issues.

Interactions with Surface Water

Although interrelated, surface water and groundwater have often been studied and managed as separate resources. Surface water seeps through the soil and becomes groundwater. Conversely, groundwater can also feed surface water sources. Sources of surface water pollution are generally grouped into two categories based on their origin.

Interactions between groundwater and surface water are complex. Consequently, groundwater pollution, sometimes referred to as groundwater contamination, is not as easily classified as surface water pollution. By its very nature, groundwater aquifers are susceptible to contamination from sources that may not directly affect surface water bodies, and the distinction of point vs. non-point source may be irrelevant. A spill or ongoing release of chemical or radionuclide contaminants into soil (located away from a surface water body) may not create point or non-point source pollution but can contaminate the aquifer below, creating a toxic plume.

Prevention

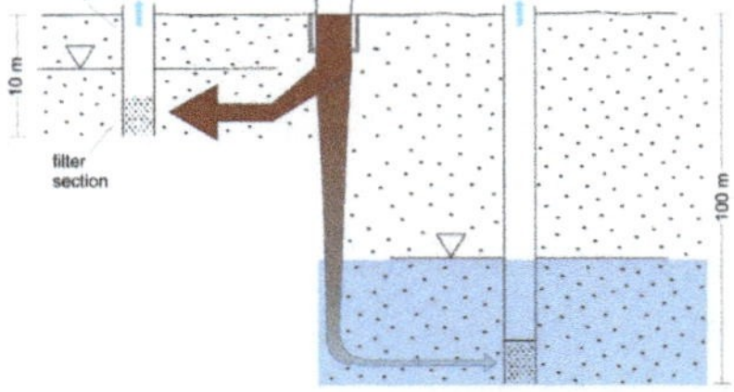

Schematic showing that there is a lower risk of groundwater pollution with greater depth of the water well

Precautionary Principle

The precautionary principle, evolved from Principle 15 of the Rio Declaration on Environment and Development, is important in protecting groundwater resources from pollution. The precautionary principle provides that "*where there are threats of irreversible damage, lack of full scientific certainty shall not be used as reason for postponing cost-effective measures to prevent environmental degradation.*".

One of the six basic principles of the European Union (EU) water policy is the application of the precautionary principle.

Groundwater Quality Monitoring

Groundwater quality monitoring programs have been implemented regularly in many countries around the world. They are important components to understand the hydrogeological system, and for the development of conceptual models and aquifer vulnerability maps.

Groundwater quality must be regularly monitored across the aquifer to determine trends. Effective groundwater monitoring should be driven by a specific objective, for example, a specific contaminant of concern. Contaminant levels can be compared to the World HealthOrganization (WHO) guidelines for drinking-water quality. It is not rare that limits of contaminants are reduced as more medical experience is gained.

Sufficient investment should be given to continue monitoring over the long term. When a problem is found, action should be taken to correct it. Waterborne outbreaks in the United States decreased with the introduction of more stringent monitoring (and treatment) requirements in the early 90s.

The community can also help monitor the groundwater quality.

Land Zoning for Groundwater Protection

The development of land-use zoning maps has been implemented by several water authorities at different scales around the world. There are two types of zoning maps: aquifer vulnerability maps and source protection maps.

Aquifer Vulnerability Map

It refers to the intrinsic (or natural) vulnerability of a groundwater system to pollution. Intrinsically, some aquifers are more vulnerable to pollution than other aquifers. Shallow unconfined aquifers are more at risk of pollution because there are fewer layers to filter out contaminants.

The unsaturated zone can play an important role in retarding (and in some cases eliminating) pathogens and so must be considered when assessing aquifer vulnerability. The biological activity is greatest in the top soil layers where the attenuation of pathogens is generally most effective.

Preparation of the vulnerability maps typically involves overlaying several thematic maps of physical factors that have been selected to describe the aquifer vulnerability. The index-based parametric mapping method GOD developed by Foster and Hirata (1988) uses three generally available or readily estimated parameters, the degree of Groundwater hydraulic confinement, geological nature of the Overlying strata and Depth to groundwater. A further approach developed by the US EPA named DRASTIC employs seven hydrogeological factors to develop an index of vulnerability: Depth to water table, net Recharge, Aquifer media, Soil media, Topography (slope), Impact on the vadose zone, and hydraulic Conductivity.

There is a particular debate among hydrogeologist whether aquifer vulnerability should be established in a general (intrinsic) way for all contaminants, or specifically for each pollutant.

Source Protection Map

It refers to the capture areas around an individual groundwater source, such as a water well or a spring, to especially protect them from pollution. Thus, potential sources of degradable pollutants, such as pathogens, can be located at distances which travel times along the flowpaths are long enough for the pollutant to be eliminated through filtration or adsorption.

Analytical methods using equations to define groundwater flow and contaminant transport are the most widely used. The WHPA is a semi-analytical groundwater flow simulation program developed by the US EPA for delineating capture zones in a wellhead protection area.

The simplest form of zoning employs fixed-distance methods where activities are excluded within a uniformly applied specified distance around abstraction points.

Locating on-site Sanitation Systems

As the health effects of most toxic chemicals arise after prolonged exposure, risk to health from chemicals is generally lower than that from pathogens. Thus, the quality of the source protection measures is an important component in controlling whether pathogens may be present in the final drinking-water.

On-site sanitation systems can be designed in such a way that groundwater pollution from these sanitation systems is prevented from occurring. Detailed guidelines have been developed to estimate safe distances to protect groundwater sources from pollution from on-site sanitation. The following criteria have been proposed for safe siting (i.e. deciding on the location) of on-site sanitation systems:

- Horizontal distance between the drinking water source and the sanitation system
    - Guideline values for horizontal separation distances between on-site sanitation systems and water sources vary widely (e.g. 15 to 100 m horizontal distance between pit latrine and groundwater wells)
- Vertical distance between drinking water well and sanitation system
- Aquifer type
- Groundwater flow direction
- Impermeable layers
- Slope and surface drainage

- Volume of leaking wastewater
- Superposition, i.e. the need to consider a larger planning area

As a very general guideline it is recommended that the bottom of the pit should be at least 2 m above groundwater level, and a minimum horizontal distance of 30 m between a pit and a water source is normally recommended to limit exposure to microbial contamination. However, no general statement should be made regarding the minimum lateral separation distances required to prevent contamination of a well from a pit latrine. For example, even 50 m lateral separation distance might not be sufficient in a strongly karstified system with a downgradient supply well or spring, while 10 m lateral separation distance is completely sufficient if there is a well developed clay cover layer and the annular space of the groundwater well is well sealed.

Legislation

Institutional and legal issues are critical in determining the success or failure of groundwater protection policies and strategies.

Sign near Mannheim, Germany indicating a zone as a dedicated "groundwater protection zone"

United States

In November 2006, the Environmental Protection Agency published the Ground Water Rule in the United States Federal Register. The EPA was worried that the ground water system would be vulnerable to contamination from fecal matter. The point of the rule was to keep microbial pathogens out of public water sources. The 2006 Ground Water Rule was an amendment of the 1996 Safe Drinking Water Act.

The ways to deal with groundwater pollution that has already occurred can be grouped into the following categories: containing the pollutants to prevent them from migrating further; removing the pollutants from the aquifer; remediating the aquifer by either immobilizing or detoxifying the contaminants while they are still in the aquifer (in-situ); treating the groundwater at its point of use; or abandoning the use of this aquifer's groundwater and finding an alternative source of water.

Management

Point-of-use Treatment

Portable water purification devices or "point-of-use" (POU) water treatment systems and field water disinfection techniques can be used to remove some forms of groundwater pollution prior to drinking, namely any fecal pollution. Many commercial portable water purification systems or chemical additives are available which can remove pathogens, chlorine, bad taste, odors, and heavy metals like lead and mercury.

Techniques include boiling, filtration, activated charcoal absorption, chemical disinfection, ultraviolet purification, ozone water disinfection, solar water disinfection, solar distillation, homemade water filters.

Arsenic removal filters (ARF) are dedicated technologies typically installed to remove arsenic. Many of these technologies require a capital investment and long-term maintenance. Filters in Bangladesh are usually abandoned by the users due to their high cost and complicated maintenance, which is also quite expensive.

Groundwater Remediation

Groundwater pollution is much more difficult to abate than surface pollution because groundwater can move great distances through unseen aquifers. Non-porous aquifers such as clays partially purify water of bacteria by simple filtration (adsorption and absorption), dilution, and, in some cases, chemical reactions and biological activity; however, in some cases, the pollutants merely transform to soil contaminants. Groundwater that moves through open fractures and caverns is not filtered and can be transported as easily as surface water. In fact, this can be aggravated by the human tendency to use natural sinkholes as dumps in areas of karst topography.

Pollutants and contaminants can be removed from ground water by applying various techniques thereby making it safe for use. Ground water treatment (or remediation) techniques span biological, chemical, and physical treatment technologies. Most ground water treatment techniques utilize a combination of technologies. Some of the biological treatment techniques include bioaugmentation, bioventing, biosparging, bioslurping, and phytoremediation. Some chemical treatment techniques include ozone and oxygen gas injection, chemical precipitation, membrane separation, ion exchange, carbon absorption, aqueous chemical oxidation, and surfactant enhanced recovery. Some chemical techniques may be implemented using nanomaterials. Physical treatment techniques include, but are not limited to, pump and treat, air sparging, and dual phase extraction.

Abandonment

If treatment or remediation of the polluted groundwater is deemed to be too difficult or expensive then abandoning the use of this aquifer's groundwater and finding an alternative source of water is the only other option.

Society and Culture

Hinkley, U.S.

The town of Hinkley, California (U.S.), had its groundwater contaminated with hexavalent chromium starting in 1952, resulting in a legal case against Pacific Gas & Electric (PG&E) and a multimillion-dollar settlement in 1996. The legal case was dramatized in the film *Erin Brockovich,* released in 2000.

California, U.S.

Nitrates and water contamination in California's Central Valley

Walkerton, Canada

In the year 2000, groundwater pollution occurred in the small town of Walkerton, Canada leading to seven deaths in what is known as the Walkerton *E. Coli* outbreak. The water supply which was drawn from groundwater became contaminated with the highly dangerous O157:H7 strain of *E. coli* bacteria. This contamination was due to farm runoff into an adjacent water well that was vulnerable to groundwater pollution.

Lusaka, Zambia

The peri-urban areas of Lusaka, the capital of Zambia, have ground conditions which are strongly karstified and for this reason – together with the increasing population density in these peri-urban areas – pollution of water wells from pit latrines is a major public health threat there.

Detailed Classification of Water Pollutants

The various types of water pollutants can be classified in to following major categories: (1) Organicpollutants, (2) Pathogens, (3) Nutrients and agriculture runoff, (4) Suspended solids and sediments (organic and inorganic), (5) Inorganic pollutants (salts and metals), (6) Thermal Pollution, and (7) Radioactive pollutants.

Organic Pollutants

Organic pollutants can be further divided into following categories:

a) Oxygen Demanding Wastes

The wastewaters such as, domestic and municipal sewage, wastewater from food processing industries, canning industries, slaughter houses, paper and pulp mills, tanneries, breweries, distilleries, etc. have considerable concentration of biodegradable organic compounds either in suspended, colloidal or dissolved form. These wastes undergo degradation and decomposition by bacterial activity.

The dissolved oxygen available in the water body will be consumed for aerobic oxidation of organic matter present in the wastewater. Hence, depletion of the DO will be a serious problem adversely affecting aquatic life, if the DO falls below 4.0 mg/L. This decrease of DO is an index of pollution.

b) Synthetic Organic Compounds

Synthetic organic compounds are also likely to enter the ecosystem through various manmade activities such as production of these compounds, spillage during transportation, and their uses in different applications. These include synthetic pesticides, synthetic detergents, food additives, pharmaceuticals, insecticides, paints, synthetic fibers, plastics, solvents and volatile organic compounds (VOCs). Most of these compounds are toxic and biorefractory organics i.e., they are resistant to microbial degradation.

Even concentration of some of these in traces may make water unfit for different uses. The detergents can form foams and volatile substances may cause explosion in sewers. Polychlorinated biphenyls (PCBs) are used in the industries since 1930s which are complex mixtures of chlorobiphenyls. Being a fat soluble they move readily through the environment and within the tissues or cells. Once introduced into environment, these compounds are exceedingly persistent and their stability to chemical reagents is also high.

c) Oil

Oil is a natural product which results from the plant remains fossilized over millions of years, under marine conditions. It is a complex mixture of hydrocarbons and degradable under bacterial action, the biodegradation rate is different for different oils, tars being one of the slowest. Oil enters in to water through oil spills, leak from oil pipes, and wastewater from production and refineries.

Being lighter than water it spreads over the surface of water, separating the contact of water with air, hence resulting in reduction of DO. This pollutant is also responsible for endangering water birds and coastal plants due to coating of oils and adversely affecting the normal activities. It also results in reduction of light transmission through surface waters, thereby reducing the photosynthetic activity of the aquatic plants. Oil includes polycyclic aromatic hydrocarbons (PAH), some of which are known to be carcinogenic.

Pathogens

The pathogenic microorganisms enter in to water body through sewage discharge as a major source or through the wastewater from industries like slaughterhouses. Viruses and bacteria can cause water borne diseases, such as cholera, typhoid, dysentery, polio and infectious hepatitis in human.

Nutrients

The agriculture run-off, wastewater from fertilizer industry and sewage contains substantial concentration of nutrients like nitrogen and phosphorous. These waters supply nutrients to the plants and may stimulate the growth of algae and other aquatic weeds in receiving waters. Thus, the value of the water body is degraded. In long run, water body reduces DO, leads to eutrophication and ends up as a dead pool of water. People swimming in eutrophic waters containing blue-green algae can have skin and eye irritation, gastroenteritis and vomiting. High nitrogen levels in the water supply, causes a potential risk, especially to infants under six months. This is when the methemoglobin results in a decrease in the oxygen carrying capacity of the blood (blue baby disease) as nitrate ions in the blood readily oxidize ferrous ions in the hemoglobin.

In freshwater systems, eutrophication is a process whereby water bodies receive excess inorganic nutrients, especially N and P, which stimulate excessive growth of plants and algae. Eutrophication can happen naturally in the normal succession of some freshwater ecosystems. However, when the nutrient enrichment is due to the activities of humans, sometimes referred to as "cultural eutrophication", the rate of this natural process is greatly intensified. Two major nutrients, nitrogen (N) and phosphorus (P), occur in streams in various forms as ions or dissolved in solution. Aquatic plants convert dissolved inorganic forms of nitrogen (nitrate, nitrite, and ammonium) and phosphorus (orthophosphate) into organic or particulate forms for use in higher trophic production. The main effects caused by eutrophication can be summarized as follows:

1. Species diversity decreases and the dominant biota changes
2. Plant and animal biomass increase
3. Turbidity increases
4. Rate of sedimentation increases, shortening the lifespan of the lake, and
5. Anoxic conditions may develop.

Suspended Solids and Sediments

These comprise of silt, sand and minerals eroded from land. These appear in the water through the surface runoff during rainy season and through municipal sewers. This can lead to the siltation, reduces storage capacities of reservoirs. Presence of suspended solids can block the sunlight penetration in the water, which is required for the photosynthesis by bottom vegetation. Deposition of the solids in the quiescent stretches of the stream or ocean bottom can impair the normal aquatic life and affect the diversity of the aquatic ecosystem. If the deposited solids are organic in nature, they will undergo decomposition leading to development of anaerobic conditions. Finer suspended solids such as silt and coal dust may injure the gills of fishes and cause asphyxiation.

Inorganic Pollutants

Apart from the organic matter discharged in the water body through sewage and industrial wastes, high concentration of heavy metals and other inorganic pollutants contaminate the water. These compounds are non-biodegradable and persist in the environment. These pollutants include mineral acids, inorganic salts, trace elements, metals, metals compounds, complexes of metals with organic compounds, cyanides, sulphates, etc.

- The accumulation of heavy metals may have adverse effect on aquatic flora and fauna and may constitute a public health problem where contaminated organisms are used for food.
- Algal growth due to nitrogen and phosphorous compounds can be observed.
- Metals in high concentration can be toxic to biota e.g. Hg, Cu, Cd, Pb, As, and Se. Copper greater than 0.1 mg/L is toxic to microbes.

Thermal Pollution

Considerable thermal pollution results due to discharge of hot water from thermal power plants, nuclear power plants, and industries where water is used as coolant. As a result of hot water discharge, the temperature of water body increases. Rise in temperature reduces the DO content of the water, affecting adversely the aquatic life. This alters the spectrum of organisms, which can adopt to live at that temperature and DO level. When organic matter is also present, the bacterial action increases due to rise in temperature; hence, resulting in rapid decrease of DO. The discharge of hot water leads to the thermal stratification in the water body, where hot water will remain on the top.

Radioactive Pollutants

Radioactive materials originate from the following:

- Mining and processing of ores,
- Use in research, agriculture, medical and industrial activities, such as I^{131}, P^{32}, Co^{60}, Ca^{45}, S^{35}, C^{14}, etc.
- Radioactive discharge from nuclear power plants and nuclear reactors, e.g., Sr^{90}, Cesium Cs^{137}, Plutonium Pu^{248}, Uranium-238, Uranium-235,
- Uses and testing of nuclear weapons.

These isotopes are toxic to the life forms; they accumulate in the bones, teeth and can cause serious disorders. The safe concentration for lifetime consumption is 1×10^{-7} microcuries per ml.

Water Quality

A rosette sampler is used to collect water samples in deep water, such as the Great Lakes or oceans, for water quality testing.

Water quality refers to the chemical, physical, biological, and radiological characteristics of water. It is a measure of the condition of water relative to the requirements of one or more biotic species and or to any human need or purpose. It is most frequently used by reference to a set of standards against which compliance can be assessed. The most common standards used to assess water quality relate to health of ecosystems, safety of human contact, and drinking water.

Standards

In the setting of standards, agencies make political and technical/scientific decisions about how the water will be used. In the case of natural water bodies, they also make some reasonable estimate of pristine conditions. Natural water bodies will vary in response to environmental conditions. Environmental scientists work to understand how these systems function, which in turn helps to identify the sources and fates of contaminants. Environmental lawyers and policymakers work to define legislation with the intention that water is maintained at an appropriate quality for its identified use.

The vast majority of surface water on the Earth is neither potable nor toxic. This remains true when seawater in the oceans (which is too salty to drink) is not counted. Another general perception of water quality is that of a simple property that tells whether water is polluted or not. In fact, water quality is a complex subject, in part because water is a complex medium intrinsically tied to the ecology of the Earth. Industrial and commercial activities (e.g. manufacturing, mining, construction, transport) are a major cause of water pollution as are runoff from agricultural areas, urban runoff and discharge of treated and untreated sewage.

Categories

The parameters for water quality are determined by the intended use. Work in the area of water quality tends to be focused on water that is treated for human consumption, industrial use, or in the environment.

Human Consumption

Contaminants that may be in untreated water include microorganisms such as viruses, protozoa and bacteria; inorganic contaminants such as salts and metals; organic chemical contaminants from industrial processes and petroleum use; pesticides and herbicides; and radioactive contaminants. Water quality depends on the local geology and ecosystem, as well as human uses such as sewage dispersion, industrial pollution, use of water bodies as a heat sink, and overuse (which may lower the level of the water).

The United States Environmental Protection Agency (EPA) limits the amounts of certain contaminants in tap water provided by US public water systems. The Safe Drinking Water Act authorizes EPA to issue two types of standards:

- *primary standards* regulate substances that potentially affect human health;
- *secondary standards* prescribe aesthetic qualities, those that affect taste, odor, or appearance.

The U.S. Food and Drug Administration (FDA) regulations establish limits for contaminants in bottled water that must provide the same protection for public health. Drinking water, including bottled water, may reasonably be expected to contain at least small amounts of some contaminants. The presence of these contaminants does not necessarily indicate that the water poses a health risk.

In urbanized areas around the world, water purification technology is used in municipal water systems to remove contaminants from the source water (surface water or groundwater) before it is distributed to homes, businesses, schools and other recipients. Water drawn directly from a stream, lake, or aquifer and that has no treatment will be of uncertain quality.

Industrial and Domestic use

Dissolved minerals may affect suitability of water for a range of industrial and domestic purposes. The most familiar of these is probably the presence of ions of calcium (Ca^{2+}) and magnesium (Mg^{2+}) which interfere with the cleaning action of soap, and can form hard sulfate and soft carbonatedeposits in water heaters or boilers. Hard water may be softened to remove these ions. The softening process often substitutes sodiumcations. Hard water may be preferable to soft water for human consumption, since health problems have been associated with excess sodium and with calcium and magnesium deficiencies. Softening decreases nutrition and may increase cleaning effectiveness. Various industries' wastes and effluents can also pollute the water quality in receiving bodies of water.

Environmental Water Quality

Environmental water quality, also called ambient water quality, relates to water bodies such as lakes, rivers, and oceans. Water quality standards for surface waters vary significantly due to different environmental conditions, ecosystems, and intended human uses. Toxic substances and high populations of certain microorganisms can present a health hazard for non-drinking purposes such as irrigation, swimming, fishing, rafting, boating, and industrial uses. These conditions may also affect wildlife, which use the water for drinking or as a habitat. Modern water quality laws generally specify protection of fisheries and recreational use and require, as a minimum, retention of current quality standards.

Urban runoff discharging to coastal waters

Satirical cartoon by William Heath, showing a woman observing monsters in a drop of London water (at the time of the *Commission on the London Water Supply* report, 1828)

There is some desire among the public to return water bodies to pristine, or pre-industrial conditions. Most current environmental laws focus on the designation of particular uses of a water body. In some countries these designations allow for some water contamination as long as the particular type of contamination is not harmful to the designated uses. Given the landscape changes (e.g., land development, urbanization, clearcutting in forested areas) in the watersheds of many freshwater bodies, returning to pristine conditions would be a significant challenge. In these cases, environmental scientists focus on achieving goals for maintaining healthy ecosystems and may concentrate on the protection of populations of endangered species and protecting human health.

Sampling and Measurement

The complexity of water quality as a subject is reflected in the many types of measurements of water quality indicators. The most accurate measurements of water quality are made on-site, because water exists in equilibrium with its surroundings. Measurements commonly made on-site and in direct contact with the water source in question include temperature, pH, dissolved oxygen, conductivity, oxygen reduction potential (ORP), turbidity, and Secchi disk depth.

Sample Collection

An automated sampling station installed along the East Branch Milwaukee River, New Fane, Wisconsin. The cover of the 24-bottle autosampler (center) is partially raised, showing the sample bottles inside. The autosampler was programmed to collect samples at time intervals, or proportionate to flow over a specified period. The data logger (white cabinet) recorded temperature, specific conductance, and dissolved oxygen levels.

More complex measurements are often made in a laboratory requiring a water sample to be collected, preserved, transported, and analyzed at another location. The process of water sampling introduces two significant problems:

- The first problem is the extent to which the sample may be representative of the water source of interest. Many water sources vary with time and with location. The measurement of interest may vary seasonally or from day to night or in response to some activity of man or natural populations of aquatic plants and animals. The measurement of interest may vary with distances from the water boundary with overlying atmosphere and underlying or confining soil. The sampler must determine if a single time and location meets the needs of the investigation, or if the water use of interest can be satisfactorily assessed by averaged values with time and/or location, or if critical maxima and minima require individual measurements over a range of times, locations and/or events. The sample collection procedure must assure correct weighting of individual sampling times and locations where averaging is appropriate. Where critical maximum or minimum values exist, statistical methods must be applied to observed variation to determine an adequate number of samples to assess probability of exceeding those critical values.

- The second problem occurs as the sample is removed from the water source and begins to establish chemical equilibrium with its new surroundings – the sample container. Sample containers must be made of materials with minimal reactivity with substances to be measured; and pre-cleaning of sample containers is important. The water sample may dissolve part of the sample container and any residue on that container, or chemicals dissolved in the water sample may sorb onto the sample container and remain there when the water is poured out for analysis.[:4] Similar physical and chemical interactions may take place with any pumps, piping, or intermediate devices used to transfer the water sample into the sample container. Water collected from depths below the surface will normally be held at the reduced pressure of the atmosphere; so gas dissolved in the water may escape into unfilled space at the top of the container. Atmospheric gas present in that air space may also dissolve into the water sample. Other chemical reaction equilibria may change if the water sample changes temperature. Finely divided solid particles formerly suspended by water turbulence may settle to the bottom of the sample container, or a solid phase may form from biological growth or chemical precipitation. Microorganisms within the water sample may biochemically alter concentrations of oxygen, carbon dioxide, and organic compounds. Changing carbon dioxide concentrations may alter pH and change solubility of chemicals of interest. These problems are of special concern during measurement of chemicals assumed to be significant at very low concentrations.

Sample preservation may partially resolve the second problem. A common procedure is keeping samples cold to slow the rate of chemical reactions and phase change, and analyzing the sample as soon as possible; but this merely minimizes the changes rather than preventing them. A useful procedure for determining influence of sample containers during delay between sample collection and analysis involves preparation for two artificial samples in advance of the sampling event. One sample container is filled with water known from previous analysis to contain no detectable amount of the chemical of interest. This sample, called a "blank", is opened for exposure to the atmosphere when the sample of interest is collected, then resealed and transported to the laboratory with the sample for analysis to determine if sample holding procedures introduced any measurable amount of the chemical of interest. The second artificial sample is collected with the sample of interest, but then "spiked" with a measured additional amount of the chemical of interest at the time of collection. The blank and spiked samples are carried with the sample of interest and analyzed by the same methods at the same times to determine any changes indicating gains or losses during the elapsed time between collection and analysis.

Filtering a manually collected water sample (grab sample) for analysis

Testing in Response to Natural Disasters and other Emergencies

Inevitably after events such as earthquakes and tsunamis, there is an immediate response by the aid agencies as relief operations get underway to try and restore basic infrastructure and provide the basic fundamental items that are necessary for survival and subsequent recovery. Access to clean drinking water and adequate sanitation is a priority at times like this. The threat of disease increases hugely due to the large numbers of people living close together, often in squalid conditions, and without proper sanitation.

After a natural disaster, as far as water quality testing is concerned there are widespread views on the best course of action to take and a variety of methods can be employed. The key basic water quality parameters that need to be addressed in an emergency are bacteriological indicators of fecal contamination, free chlorine residual, pH, turbidity and possibly conductivity/total dissolved solids. There are a number of portable water test kits on the market widely used by aid and relief agencies for carrying out such testing.

After major natural disasters, a considerable length of time might pass before water quality returns to pre-disaster levels. For example, following the 2004 Indian Ocean tsunami the Colombo-based International Water Management Institute (IWMI) monitored the effects of saltwater and concluded that the wells recovered to pre-tsunami drinking water quality one and a half years after the event. IWMI developed protocols for cleaning wells contaminated by saltwater; these were subsequently officially endorsed by the World Health Organization as part of its series of Emergency Guidelines.

Chemical Analysis

The simplest methods of chemical analysis are those measuring chemical elements without respect to their form. Elemental analysis for oxygen, as an example, would indicate a concentration of 890,000 milligrams per litre (mg/L) of water sample because water is made of oxygen. The method selected to measure dissolved oxygen should differentiate between diatomic oxygen and oxygen combined with other elements. The comparative simplicity of elemental analysis has produced a large amount of sample data and water quality criteria for elements sometimes identified as heavy metals. Water analysis for heavy metals must consider soil particles suspended in the water sample. These suspended soil particles may contain measurable amounts of metal. Although the particles are not dissolved in the water, they may be consumed by people drinking the water. Adding acid to a water sample to prevent loss of dissolved metals onto the sample container may dissolve more metals from suspended soil particles. Filtration of soil particles from the water sample before acid addition, however, may cause loss of dissolved metals onto the filter. The complexities of differentiating similar organic molecules are even more challenging.

A gas chromatograph mass spectrometer measures pesticides and other organic pollutants

Making these complex measurements can be expensive. Because direct measurements of water quality can be expensive, ongoing monitoring programs are typically conducted by government agencies. However, there are local volunteer programs and resources available for some general assessment. Tools available to the general public include on-site test kits, commonly used for home fish tanks, and biological assessment procedures.

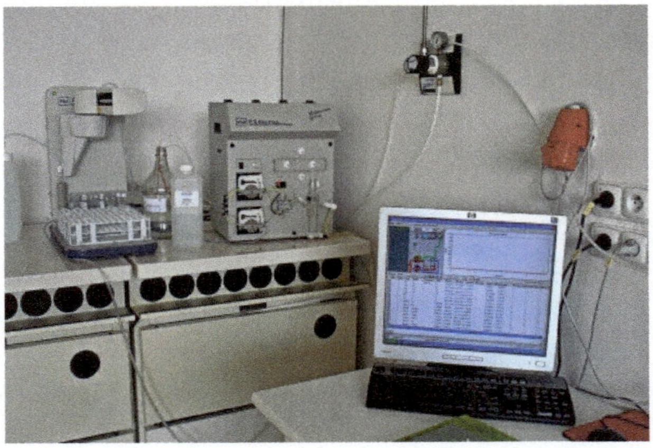

Atomic fluorescence spectroscopy is used to measure mercury and other heavy metals

Real-time Monitoring

Although water quality is usually sampled and analyzed at laboratories, nowadays, citizens demand real-time information about the water they are drinking. During the last years, several companies are deploying worldwide real-time remote monitoring systems for measuring water pH, turbidity or dissolved oxygen levels.

Drinking Water Indicators

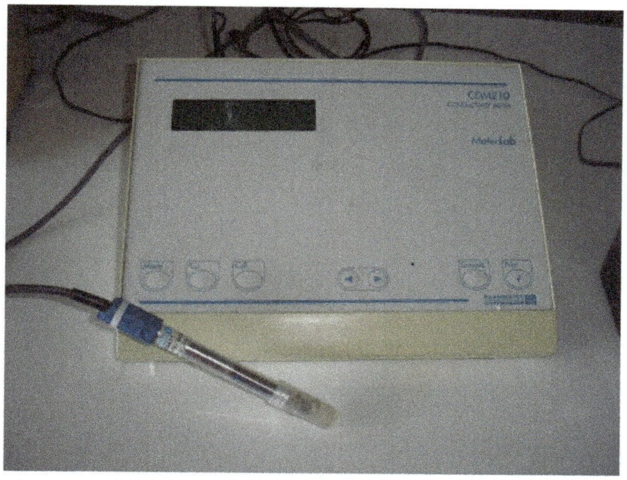

An electrical conductivity meter is used to measure total dissolved solids

The following is a list of indicators often measured by situational category:

- Alkalinity
- Color of water
- pH
- Taste and odor (geosmin, 2-Methylisoborneol (MIB), etc.)
- Dissolved metals and salts (sodium, chloride, potassium, calcium, manganese, magnesium)
- Microorganisms such as fecal coliform bacteria (*Escherichia coli*), Cryptosporidium, and Giardia lamblia
- Dissolved metals and metalloids (lead, mercury, arsenic, etc.)
- Dissolved organics: colored dissolved organic matter (CDOM), dissolved organic carbon (DOC)
- Radon
- Heavy metals
- Pharmaceuticals
- Hormone analogs

Environmental Indicators

Physical Indicators

• Water temperature	• Total dissolved solids (TDS)
• Specific conductance or electrical conductance (EC) or conductivity	• Odour of water
• Total suspended solids (TSS)	• Color of water
• Transparency or turbidity	• Taste of water

Chemical Indicators

• pH	• Heavy metals
• Biochemical oxygen demand (BOD)	• Nitrate
• Chemical oxygen demand (COD)	• Orthophosphates
• Dissolved oxygen (DO)	• Pesticides
• Total hardness (TH)	• Surfactants

Biological indicators

• Ephemeroptera	• *Escherichia coli* (E. coli)
• Plecoptera	• Coliform bacteria
• Mollusca	
• Trichoptera	

Biological monitoring metrics have been developed in many places, and one widely used measure is the presence and abundance of members of the insect orders Ephemeroptera, Plecoptera and Trichoptera (common names are, respectively, mayfly, stonefly and caddisfly). EPT indexes will naturally vary from region to region, but generally, within a region, the greater the number of taxa from these orders, the better the water quality. Organisations in the United States, such as EPA offer guidance on developing a monitoring program and identifying members of these and other aquatic insect orders.

Individuals interested in monitoring water quality who cannot afford or manage lab scale analysis can also use biological indicators to get a general reading of water quality. One example is the IOWATER volunteer water monitoring program, which includes a benthicmacroinvertebrate indicator key.

Bivalve molluscs are largely used as bioindicators to monitor the health of aquatic environments in both fresh water and the marine environments. Their population status or structure, physiology, behaviour or the level of contamination with elements or compounds can indicate the state of contamination status of the ecosystem. They are particularly useful since they are sessile so that they are representative of the environment where they are sampled or placed. A typical project is the Mussel Watch Programme, but today they are used worldwide.

The Southern African Scoring System (SASS) method is a biological water quality monitoring system based on the presence of benthic macroinvertebrates. The SASS aquatic biomonitoring tool has been refined over the past 30 years and is now on the fifth version (SASS5) which has been specifically modified in accordance with international standards, namely the ISO/IEC 17025 protocol. The SASS5 method is used by the South African Department of Water Affairs as a standard method for River Health Assessment, which feeds the national River Health Programme and the national Rivers Database.

Standards and Reports

International

- The World Health Organisation (WHO) has published guidelines for drinking-water quality (GDWQ) in 2011.
- The International Organization for Standardization (ISO) published regulation of water quality in the section of ICS 13.060, ranging from water sampling, drinking water, industrial class water, sewage, and examination of water for chemical, physical or biological properties. ICS 91.140.60 covers the standards of water supply systems.

National Specifications for Ambient Water and Drinking Water

European Union

The water policy of the European Union is primarily codified in three directives:

- Directive on Urban Waste Water Treatment (91/271/EEC) of 21 May 1991 concerning discharges of municipal and some industrial wastewaters;
- The Drinking Water Directive (98/83/EC) of 3 November 1998 concerning potable water quality;
- Water Framework Directive (2000/60/EC) of 23 October 2000 concerning water resources management.

India

- Indian Council of Medical Research (ICMR) Standards for Drinking Water.

South Africa

Water quality guidelines for South Africa are grouped according to potential user types (e.g. domestic, industrial) in the 1996 Water Quality Guidelines. Drinking water quality is subject to the South African National Standard (SANS) 241 Drinking Water Specification.

United Kingdom

In England and Wales acceptable levels for drinking water supply are listed in the "Water Supply (Water Quality) Regulations 2000."

United States

In the United States, Water Quality Standards are defined by state agencies for various water bodies, guided by the desired uses for the water body (e.g., fish habitat, drinking water supply, recreational use). The Clean Water Act (CWA) requires each governing jurisdiction (states, territories, and covered tribal entities) to submit a set of biennial reports on the quality of water in their area. These reports are known as the 303(d) and 305(b) reports, named for their respective CWA provisions, and are submitted to, and approved by, EPA. These reports are completed by the governing jurisdiction, typically a state environmental agency. EPA recommends that each state submit a single "Integrated Report" comprising its list of impaired waters and the status of all water bodies in the state. The *National Water Quality Inventory Report to Congress* is a general report on water quality, providing overall information about the number of miles of streams and rivers and their aggregate condition. The CWA requires states to adopt standards for each of the possible designated uses that they assign to their waters. Should evidence suggest or document that a stream, river or lake has failed to meet the water quality criteria for one or more of its designated uses, it is placed on a list of impaired waters. Once a state has placed a water body on this list, it must develop a management plan establishing Total Maximum Daily Loads (TMDLs) for the pollutant(s) impairing the use of the water. These TMDLs establish the reductions needed to fully support the designated uses.

Drinking water standards, which are applicable to public water systems, are issued by EPA under the Safe Drinking Water Act.

Surface Water Quality: Rivers and Streams

Surface water is highly susceptible to contamination due to sewage, industrial waste discharge, and agricultural run-off apart from the recreational activities. The objective of the water quality management is to control the discharge of pollutants so that water quality is not degraded to an unacceptable extent below the acceptable level. To achieve this quantitative analysis of pollutants is must. It is necessary to understand the background water quality which would be present without human interventions, and decide the levels of pollutants acceptable for intended uses of the water. The impact of different pollutants joining the water body should be understood well for this purpose. The impact of the pollution on a river depends both on the nature of the pollutant and the characteristics of the individual river. Some of the important characteristics include velocity of water flowing in the river, depth of flow, type of river bottom, and surrounding vegetation. The other factors include, climate of the region, geological characteristics of the watershed, land used pattern, and aquatic life in the rivers. All these parameters must be considered in water quality management in the river.

It is understood that the water quality in the rivers will not remain same throughout the stretch of river i.e. from origin to the point where it meet to the sea. The water quality is best near the origin of the river and goes on deteriorating as the river flows and goes on assimilating the pollutant discharged in it. Proper management is necessary if the water quality is to remain usable for intended purpose. Oxygen demanding wastes and nutrients are among the common pollutant having profound impact on almost all types of rivers, hence they deserve special emphasis.

Effect of Oxygen Demanding Wastes on Rivers

Depletion of dissolved oxygen is a major problem due discharge of oxygen demanding organic or inorganic pollutant in the surface water. This poses threat to higher forms of aquatic life, if the concentration of oxygen falls below a critical point. To quantify how much oxygen will be depleted, it is necessary to know the quantity of oxygen demanding waste and how much oxygen will be required to degrade the waste. Although, oxygen is getting depleted for the degradation of organic matter, it is continuously being replenished from the atmosphere and through photosynthesis. The net concentration of oxygen in the water body is determined by the relative rates of these competing processes i.e. deoxygenation and reoxygenation.

Estimation of Organic Content of the Wastewater

The organic matter present in the water body can be analyzed in laboratory by determining Biochemical Oxygen Demand (BOD), Chemical Oxygen Demand (COD), and by determination of Total Organic Carbon (TOC). These test procedures and relevance of these tests are discussed below.

Biochemical Oxygen Demand (BOD)

The BOD can be defined as the oxygen required for biochemical oxidation of organic matter present in the water under aerobic conditions. This test is based on the premise that all the biodegradable organic matter contained in a water sample will be oxidized to CO_2 and H_2O by microorganisms using molecular oxygen. For example, the general overall oxidation reaction for glucose is

$$C_6H_{12}O_6 + 6O_2 \longrightarrow 6CO_2 + 6H_2O$$

Thus, the theoretical oxygen demand would then be:

$$\text{Oxygen demand} = (\text{Gram of oxygen used}) / (\text{Gram of carbon oxidized})$$

$$= 192 / 72 = 2.67 \text{ g/g of carbon}$$

The actual BOD will be less than theoretical oxygen demand due to incorporation of some of the carbon into newly synthesized bacterial cells. The test is performed under the conditions similar to those in actual natural water to measure indirectly the amount of biodegradable organic matter present. A water sample is inoculated with bacteria (1 to 2 mL of sewage per liter) that consume the biodegradable organic matter to obtain energy for their life processes. The organisms also utilizes oxygen in the process of consuming the organic matter, the process is called as 'aerobic' decomposition. This oxygen consumption is measured; more is the organic matter concentration more is the amount of oxygen utilized. Thus, the BOD test is the indirect measurement of organic matter in terms of the oxygen requirement to convert them into stable end product. Although, not all organic matter is biodegradable within the stipulated incubation period, and the actual test procedure lack in precision due to different inoculum seed and many fold dilution required, it is still the most widely used method of quantifying organic matter because of the direct conceptual relationship between BOD and oxygen depletion in receiving waters.

The BOD test is performed for the following:

- To determine quantity of oxygen required forbiochemical stabilization of organic matter.
- To determine suitability of biological treatment method, depending on COD/BOD ratio, and sizing the treatment units. For COD/BOD ratio less than 2 the wastewater is most suitable for biological treatment; for COD/BOD ratio between 2 to 6 acclimation of the inoculum is required for effective biological treatment; and for higher COD/BOD ratio biological treatment of such wastewater may not be suitable option.
- To monitor efficiency of the process.
- To determine compliance with wastewater discharge permits.

During the BOD test the organic matter will be converted into stable end product such as CO_2, sulphate (SO_4), orthophosphate (PO_4) and nitrate (NO_3).

The simple representation of carbonaceous BOD can be explained as below:

$$\text{Organic matter} + O_2 \xrightarrow{\text{microorganisms}} CO_2 + H_2O + \text{New Cells} + \text{Stable products}$$

This reaction continues till sufficient DO is available in the water. When DO is not available condition becomes anaerobic decomposition (fermentative reduction). The reaction under anaerobic conditions is as under:

$$\text{Organic matter} \xrightarrow{\text{microorganisms}} CO_2 + CH_4 + \text{New Cells} + \text{Other products} \left(NH_3,\ H_2S\right)$$

BOD Test

Biochemical oxidation is slow process and theoretically takes an infinite time to go to completion i.e. complete oxidation of organic matter. During the first few days the rate of oxygen depletion is rapid because of the high concentration of organic matter present. As the concentration of organic matter decreases, so does the rate of oxygen consumption. Also initially concentration of easily biodegradable organic matter will be more and as the time proceeds fraction of this component will deplete faster as compared to total organic matter reduction. Hence, during initial days the rate of BOD exerted will be more as compared to later days. During the last part of the BOD curve, oxygen consumption is mostly associated with the decay of the bacteria that grew during the early part of the test. The oxygen consumption typically follows the pattern as shown in the figure below. For wastewater like sewage, within 20 day period, the oxidation of carbonaceous organic matter is about 95 to 99% complete,and in the first five days, the period used for BOD determination, 60 to 70% oxidation is complete. The 20 °C temperature used is an average temperature value typically for slow moving streams in temperate climate. Different results would be obtained at different temperatures because biochemical reaction rates are temperature dependent.

The biochemical oxygen demand is represented as BOD_5 20°C, which indicates the total amount of oxygen consumed for biochemical oxidation of organic matter for first five days at 20oC incubation temperature. Under Indian conditions, the BOD values are acceptable for 3 days incubation at 27 °C temperature.

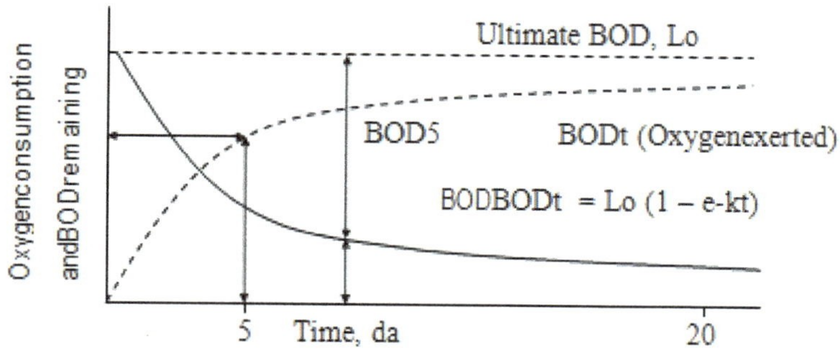

Variation in DO profile during BOD test with duration of incubation

Since, the saturation value of DO for water at 20°C is only 9.1 mg/L, it is usually necessary to dilute the samples to keep final DO level, at the end of incubation period, above 1.5 mg/L. Hence, according to BOD values expected for that wastewater appropriate dilution should be carried out. Thus, the actual BOD of the unseeded sample can be worked out as

$$\text{The 5 day BOD of sample} = \frac{DO_i - DO_f}{p} \quad (1)$$

Where, DO_i and DO_f are initial and final DO of diluted wastewater sample

$$\text{p is the dilution fraction} = \frac{\text{Volume of wastewater}}{\text{Volume of wastewater + volumedilution water}} \quad (2)$$

The total volume of the BOD bottle used for test is usually 300 mL. The dilution water (distilled water) is aerated for sufficient time to correct DO close to the saturation value. Nutrients and buffer solutions are added to the dilution water to provide nutrient for bacterial growth and maintain pH near neutral. Sufficient amount of seed is added to the BOD bottle to ensure adequate concentration of bacterial population to carry out the biodegradation.

Usually 1 to 2 mL of sewage perliteris considered as sufficient to actasaseed. In suchcaseitis necessary tos ubtract the oxygendemand of the seed from the mixed sample, because organic matter present in this 1 to 2mL of seed will also exert oxygendemand. Thus, the BOD of the wastewater with seeded sample can be worked out as below. The DO dropin blank is multiplied by (1-*p*) because this volume of blank is only present in the sample.

$$\frac{BOD_5 = (DO_i - DO_f) - (B_i - B_f)(1-p)}{p} \quad \text{...(3)}$$

Where,

DO_i an DO_f = DO of mixture, initial and final values, respectively,

B_i and B_f = DO of blank, initial and final values, respectively,

$p = V_w/V_m$ = Volume of wastewater in mixture / Total volume of mixture.

1. Dilution Water Requirement:

A wastewater is expected to have BOD_5 of about 200mg/L. The initial DO of dilution water is 8.0 mg/L. Calculate the dilution requirement for BOD determination.

Solution

BOD = 200mg/L; DOi = 8.0mg/L,

Minimum DO that should be left after five day so fincubationis 1.5–2.0mg/L, Sayfinal DO = 2.0 mg/L

Hence, dilution required=200/(8.0–2.0)=33.33 say 35 to 40 times.

Comments: To have accurate test result, the total DO drop during five days in cubation should be 2.0 mg/L.

2.BOD Determination

A test bottle containing only seeded dilution water has its DO level drop by 1.0 mg/Lina 5-day incubation. A 300 mL BOD bottle filled with 10mL of wastewater and the rest seeded dilution water experiences a DO drop of 6.2 mg/L in the same time period. What would be five day BOD of the wastewater?

Solution:

Dilution factor p=10/300

Therefore, BOD_5= [6.2–1.0(1–(10/300))]/(10/300)= 157mg/L

BOD Model

It is generally assumed that the rate at which the oxygenis consumedis directly proportional to the concentration of degra dable organic matter remaining at any time. The kinetics of BOD reaction can be formulated in accordance with first order reaction kinetics as:

$$dL_t/dt = -KL_t \quad (4)$$

Where, L_t=amount of first order BOD remaining in wastewater at time t

K=First order BOD reaction rate constant, time^{-1}

Integrating $\int_0^t dL_t = -KL_t.dt$

i.e., $[\log L_t]_0^t = -K.t$

$$Lt/Lo = e^{-K.t} \text{ or } 10^{-K.t} \quad (5)$$

Where Lo or BOD_u at time t = 0, is the ultimate first stage BOD initially present in the sample.

The relation between K (base e) and K (base 10) is

$$K\ (base10) = K\ (base\ e)/2.303 \quad (6)$$

The amount of BOD remaining at time 't' equals

$$Lt = Lo(e^{-k.t}) \quad (7)$$

The amount of BOD that has been exerted (amount of oxygen consumed) at any time 't' is given by

$$BODt = Lo - Lt = Lo(1 - e^{-k.t}) \quad (8)$$

and the five day BOD ise qual to

$$BOD_5 = Lo - L_5 = Lo(1 - e^{-5k}) \quad (9)$$

For polluted water and wastewater, a typical value of K (base e, 20^0C) is 0.23 per day and K (base 10, 20^0C) is 0.10 per day. These values vary widely for the wastewater in the range from 0.05 to 0.3 per day for K (base 10) and 0.23 to 0.7 for K (base e).

The ultimate BOD (Lo) is defined as the maximum BOD exerted by the wastewater. It is difficult to assign exact time to achieve ultimate BOD, and theoretically it takes infinite time. From the practical point of view, it can be said that when the BOD curve is approximately horizontal the ultimate BOD has been achieved. The time required to achieve the ultimate BOD depends upon the characteristics of the wastewater, i.e., chemical composition of the organic matter present in the wastewater and its biodegradable properties and temperature of incubation. At higher temperature for same concentration and nature of organic matter ultimate BOD will be achieved in shorter time as compared to lower temperatures, where it will require more time. The ultimate BOD best expresses the concentration of degradable organic matter based on the total oxygen required to oxidize it. However, it does not indicate how rapidly oxygen will be depleted in the receiving water. Oxygen depletion is related to both the ultimate BOD and the BOD rate constant (K). The ultimate BOD will increase in direct proportion to the concentration of biodegradable organic matter.

The BOD reaction rate constant is dependent on the following:

1. The nature of the waste
2. The ability of the organisms in the system to utilize the waste
3. The temperature

Nature of the waste: Thousands of organic matters exist with different chemical composition in nature. All organic matter will not have same degradation rate. Simple sugar and starches are rapidly degraded and will therefore have a high value of BOD rate constant. Cellulose degrades much more slowly and hairs are almost undegradable during BOD test or during biological treatment of wastewater. Other compounds are intermediate degradable between these extremes. For complex waste, like sewage, the BOD rate constant depends upon the relative proportions of the various components. The BOD rate constant is high for the raw sewage (K (base e) = 0.35 -0.7 per day) and low for the treated sewage (K (base e) = 0.12 - 0.23 per day), owing to the fact that, during wastewater treatment the easily biodegradable organic matter will get more completely removed than the less biodegradable organics. Hence, in the treated wastewater, relative proportion of the less biodegradable organic matter will be higher, giving lower BOD rate constant.

Ability of organisms to utilize waste: Every microorganism is limited in its ability to utilize organic compounds. Many organic matters can only be utilized by particular group of microorganisms. In natural environment, where the water course is receiving particular organic compound, the microorganisms which have capability to degrade that organic matter will grow in predominant. However, the culture used during BOD test may have very small fraction of the organisms which can degrade that particular organic compounds in the waste. As a result the BOD value, for limited incubation duration, and the rate constant would be lower in the laboratory test than in the natural water environment where the waste is regularly discharged. Therefore, the BOD test should be conducted with organisms which have been acclimated to the waste so that the rate constant determined in the laboratory can be compared to that in the river.

Temperature: The biochemical reactions are temperature dependent and the activity of the microorganism increases with the increase in temperature up to certain value, and drop with decrease in temperature. Since, the oxygen utilization in BOD test is caused by microbial metabolism, the rate of utilization is similarly affected by the temperature. The standard temperature at which BOD is determined is usually 20°C. However, the water temperature may vary from place to place for the same river; hence, the BOD rate constant is adjusted to the temperature of receiving water using following relationship:

$$K_T = K_{20}\theta^{(T-20)} \tag{10}$$

Where:

T = temperature of interest, °C

K_T = BOD rate constant at the temperature of interest, day^{-1}

K_{20} = BOD rate constant determined at 20°C, day^{-1}

θ = temperature coefficient. This has a value of 1.056 in general and 1.047 for higher temperature greater than 20°C. This is because increase in reaction rate is higher when temperature increases from 10 to 20°C as compared to when temperature is increased from 20 to 30°C.

Example:3

The treated wastewater is being discharged into a river that has a temperature of 15°C. The BOD rate constant determined in the laboratory for this mixed water is 0.12 per day. What fraction of maximum oxygen consumption will occur in first four days?

Solution

Determine the BOD rate constant at the river water temperature:

$$K_{15} = K_{20}(1.056)^{(T-20)}$$

$=0.12(1.056)^{(15-20)}$

$=0.091$ perday

Using this value of K to find the fraction of maximum oxygen consumption in four days:

$BOD_4 = Lo(1-e^{-0.091x4})$

Therefore, $BOD_4/Lo = 0.305$

Example: 4

The dissolved oxygen in an unseeded sample of diluted wastewater having an initial DO of 9.0 mg/L is measured to be 3.0 mg/L after 5 days. The dilution fraction is 0.03 and reaction rate constant k = 0.22 day^{-1}. Calculate a) 5 day BOD of the waste, b) ultimate carbonaceous BOD, and c) What would be remaining oxygen demand after 5 days?

Solution

a) Oxygen demand for first 5 days

$$BOD_5 = (DO_i - Do_f) / p = (9.0 - 3.0) / 0.03 = 200 \text{ mg/L}$$

b) Ultimate BOD

$$BOD_u = Lo = BODt / (1 - e^{-kt}) = 200 / (1 - e^{-0.22 \times 5}) = 300 \text{ mg/L}$$

c) After 5 days, 200 mg/L of oxygen demand out of total 300 mg/L would be satisfied. Hence, the remaining oxygen demand would be 300 – 200 = 100 mg/L

Example: 5

Determine ultimate BOD for a wastewater having 5 day BOD at 20°C as 160 mg/L. Assume reaction rate constant as 0.2 per day (base 10).

Solution

$$BOD_5 = Lo\,(1 - 10^{-k.t})$$

$$160 = Lo\,(1 - 10^{-5 \times 0.2})$$

Therefore, Lo = 177.8 mg/L ~ 178 mg/L

Example: 6

The BOD of a sewage incubated for one day at 30°C has been found to be 100 mg/L. What will be the five day 20°C BOD? Assume K = 0.12 (base 10) at 20°C, and θ = 1.056

Solution:

BOD at 30°C = 100 mg/L

$$K_{20} = 0.12$$

Now $K_{30} = K_{20}\,\theta^{(T-20)}$

$K_{30} = 0.12\,(1.056)^{30-20} = 0.207$ per day

$BOD_t = Lo\,(1 - 10^{-kt})$

$100 = Lo\,(1 - 10^{-0.207 \times 1})$

$Lo = 263.8$ mg/L

This is ultimate BOD, the value of which is independent of incubation temperature.

Now BOD_5 at 20°C can be calculated as:

$$BOD_5 \text{ at } 20^\circ C = Lo\,(1 - 10^{-kt}) = 263.8\,(1 - 10^{-0.12 \times 5}) = 197.5 \text{ mg/L}$$

Example: 7

Determine the 1 day BOD and ultimate first stage BOD for a wastewater whose 5 day 20°C BOD is 200 mg/L. The reaction rate constant k (base e) = 0.23 per day.

Solution

Ultimate BOD

$$BOD_u = Lo = BODt / (1 - e^{-kt}) = 200 / (1 - e^{-0.23 \times 5}) = 293 \text{ mg/L}$$

Therefore, one day BOD

$$Y1 = Lo - L1 = Lo\,(1 - e^{-kt}) = 293\,(1 - e^{-0.23 \times 1}) = 60 \text{ mg/L}$$

Interpretation of the BOD Test Result

Following factors must be considered in the interpretation of the BOD for industrial wastewaters

1. The seed is acclimated to the wastewater and lag period required for acclimation is eliminated.
2. The rate constant should be established based on long term BOD tests on both wastewater and treated effluents. The rate constant for untreated and treated wastewater is not same for many wastewaters. The rate constant value is higher for untreated wastewater and lower for treated wastewater. For example, for raw sewage rate constant is about 0.15 to 0.3 and that for treated sewage it is around 0.05 to 0.15 (base 10). Hence, direct comparison of BOD may not be valid. The value of K (base e) for raw sewage varies in the range 0.35 to 0.7 and that for treated sewage it will be 0.12 to 0.23.
3. In case of acidic waste, all samples must be neutralized before incubation.
4. When organic matter is present in suspended form, interpretation of the test result is difficult due to lag time involved in hydrolysis of organic matter before actual oxidation starts during BOD test.

Nitrification in BOD Test

Non-carbonaceous matter, such as ammonia is produced during the hydrolysis of proteins. In addition, when the living things die, excreta waste, and nitrogen organic compounds, the nitrogen tied to organic molecule is converted to ammonia by bacterial and fungal action.

Under aerobic conditions, this ammonia will be converted to nitrate, called as nitrification as per the reactions given below:

$$2NH_3 + 3O_2 \xrightarrow{\text{Nitrosomonas}} 2NO_2^- + 2H^+ + 2H_2O$$

$$2NO_2^- + O_2 \xrightarrow{\text{Nitrobacter}} 2NO_3^-$$

Hence, the organic matter containing nitrogen will have oxygen requirement for nitrification. The oxygen demand associated with the oxidation of ammonia to nitrate is called the nitrogenous BOD. Due to low growth rate of nitrifying bacteria, this BOD demand normally occurs from 6 to 10 days. This is one of the reasons to use incubation period of 5 days for BOD determination to eliminate oxygen demand for nitrification and to find out only carbonaceous oxygen demand. Incidentally, the five day period was chosen for the test because the Themes River requires five days from its origin to join sea, and if oxygen demand for these five days is determined and satisfied the river water quality can be protected.

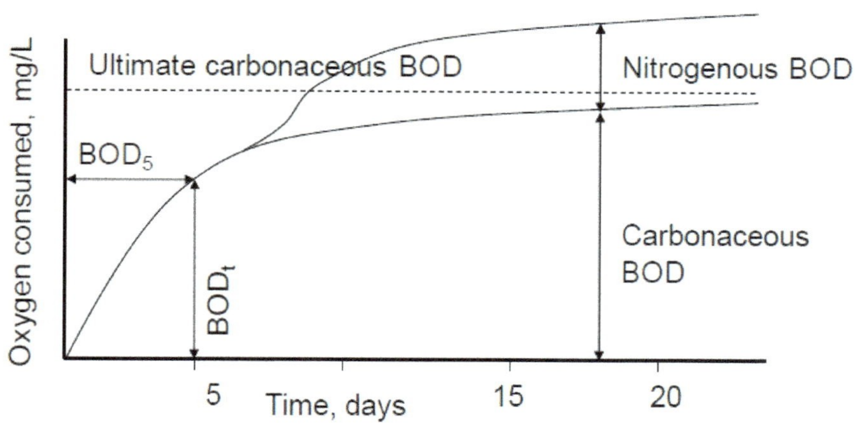

Nitrification during BOD test

Other Measures of Oxygen Demand

Chemical Oxygen Demand (COD)

During COD determination total organic content of the waste is oxidized by dichromate in acid solution.

- In this test to determine the oxygen requirement of the wastewater, strong oxidizing agent 'potassium dichromate' is used.
- Acidic environment is provided to accelerate the reactions by addition of sulphuric acid.
- The reflux flasks (or closed reflux vials), used for the test, are heated to 150°C for two hours

with silver sulphate as catalyst. When silver sulphate catalyst is used, the recovery of most organic compounds is greater than 92 percent.

- COD test measures virtually all oxidizable organic compounds whether biodegradable or not, except some aromatic compounds which resists dichromate oxidation.
- The COD is proportional to BOD only for readily soluble organic matter in dissolved form e.g. sugars.
- No correlation between BOD and COD exists when:
    - o Organic matter is present in suspended form; under such situation filtered samples should be used.
    - o Complex wastewater containing refractory substances.
- For readily biodegradable waste, such as dairy COD = BODu/0.92

The correlation between BOD and COD for sewage is presented in the Figure (Haandel and Lettinga, 1994).

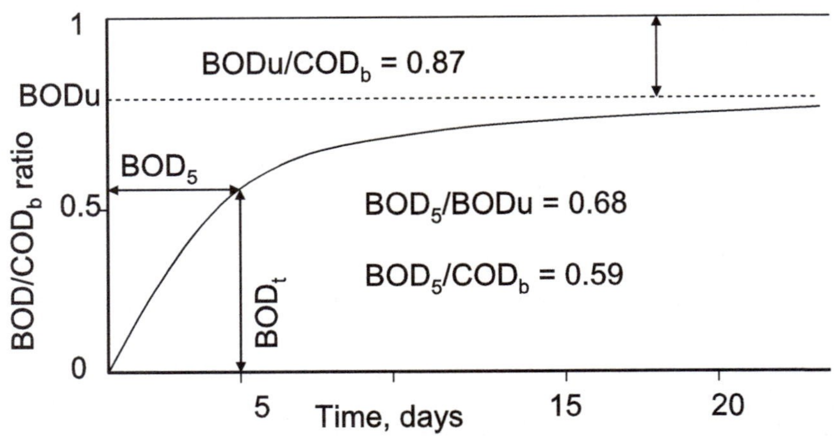

Correlation between BOD and COD for sewage at 200 C incubation and k = 0.23 per day.

The COD is faster determination but does not give idea about the nature of organic matter whether biodegradable or biorefractory organic matter. Hence, determination of BOD is necessary for the wastewater to know biodegradable organic matter fraction. The BOD is not very useful test for routine plant control due to long incubation period required, hence requiring long time (5 days) to obtain results. Thus, it is important to develop correlationbetween BOD and COD (or TOC), so that COD (or TOC) can be used as a parameter for routine analysis and control of the treatment plant. Once COD values are known, the BOD can be estimated using correlation. The test results are more reproducible for COD.

Theoretical Oxygen Demand (ThOD)

Theoretical oxygen demand for the wastewater is calculated as oxygen required for oxidizing the organic matter to end products. For example, for glucose, the theoretical oxygen demand can be worked out as below:

$$C_6H_{12}O_6+6O_2 \longrightarrow 6CO_2 +6H_2O$$

$$ThOD=(6M_{02})/(M_{C6H12O6})=(6 \times 16 \times 2)/(12 \times 6 + 1 \times 12 + 6 \times 16) = 1.07$$

For most of the organic compounds (except aromatics resisting dichromate oxidation) COD is equal to ThOD.

TOC is related to COD through carbon-oxygen balance.

$$COD/TOC= (6M_{02})/(M_C)=2.66$$

Depending on the organic matter in question COD/TOC ratio may vary from zero (for organic matter resistant to dichromate oxidation) to 5.33 for methane (most reduced organic compound). Since, organic content undergoes changes during biological oxidation, COD/TOC and BOD/TOC, BOD/COD values will change during treatment.

References

- Suthar, S; Bishnoi, P; Singh, S; et al. (2009). "Nitrate contamination in groundwater of some rural areas of Rajasthan, India". Journal of Hazardous Materials. 171 (1–3): 189–199
- Graham, J.P., Polizzotto, M.L. (2013). "Pit Latrines and Their Impacts on Groundwater Quality: A Systematic Review". Environ Health Perspect. 121 (5): 521–530. doi:10.1289/ehp.1206028. PMC 3673197. PMID 23518813
- Dickens CWS and Graham PM. 2002. The Southern Africa Scoring System (SASS) version 5 rapid bioassessment for rivers "African Journal of Aquatic Science", 27:1–10
- Smith, M; Cross, K; Paden, M; Laben, P, eds. (2016). Spring - managing groundwater sustainably (PDF). IUCN. ISBN 978-2-8317-1789-0
- "Water Quality and Technology-Based Permitting". EPA. Archived from the original on May 21, 2013. Retrieved 2013-09-29
- Singh, B; Singh, Y; Sekhon, GS (1995). "Fertilizer-N use efficiency and nitrate pollution of groundwater in developing countries". Journal of Contaminant Hydrology. 20 (3–4): 167–184. doi:10.1016/0169-7722(95)00067-4
- . Fawell, J; Bailey, K; Chilton, J; Dahi, E (2006). Fluoride in drinking-water (PDF). Geneva: IWA for WHO. ISBN 9241563192
- Huang, Yee. "Why the EPA May Be About to Take Big Steps to Clean Up Our Water". AlterNet. Retrieved April 23, 2011
- Jackson, LE; Burger, M; Cavagnaro, TR (2008). "Roots, Nitrogen Transformations, and Ecosystem Services". Annual Review of Plant Biology. 59 (1): 341–363. doi:10.1146/annurev.arplant.59.032607.092932.
- Ross, N, ed. (2010). Clearing the waters a focus on water quality solutions. Nairobi, Kenya: UNEP. ISBN 978-92-807-3074-6
- "Report Offers First Worldwide Estimate of Investments in Combating Water Pollution". American Association for the Advancement of Science. Retrieved April 18, 2011
- King, Dennis (2005). "Crunch Time for Water Quality Trading". Choices. 20 (1): 70–75. Retrieved April 25, 2011
- Dickens CWS and Graham PM. 2002. The Southern Africa Scoring System (SASS) version 5 rapid bioassessment for rivers "African Journal of Aquatic Science", 27:1–10
- Khan, MN; Mohammad, F (2014). "Eutrophication: Challenges and Solutions". In Ansari, AA; Gill, SS. Eutrophication: Causes, Consequences and Control. Springer. ISBN 978-94-007-7813-9

- "Project Development and BECC Certification". Border Environmental Cooperation Commission. Retrieved 2013-09-29
- Knobeloch, L; Salna,B; Hogan,A; Postle, J; Anderson, H (2000). "Blue Babies and Nitrate-Contaminated Well Water". Environ. Health Perspect. 108 (7): 675–8. doi:10.1289/ehp.00108675. PMC 1638204 . PMID 10903623
- Linsley, Ray K. & Franzini, Joseph B. Water-Resources Engineering (1972) McGraw-Hill ISBN 0-07-037959-9 pp.454–456
- ATSDR (US Agency for Toxic Substance & Disease Registry) (2008). "Follow-up Health Consultation: Anniston Army Depot." (PDF). Retrieved 18 March 2017
- Fewtrell,L (2004). "Drinking-Water Nitrate, Methemoglobinemia, and Global Burden of Disease: A Discussion". Environmental Health Perspectives. 112 (14): 1371–1374. doi:10.1289/ehp.7216. PMC 1247562 . PMID 15471727
- State Coalition for Remediation of Drycleaners (2011). "A Citizen's Guide to Drycleaner Cleanup" (PDF). Retrieved 20 March 2017
- International Water Management Institute, Colombo, Sri Lanka (2010). "Helping restore the quality of drinking water after the tsunami." Success Stories. Issue 7. doi:10.5337/2011.0030

3

Sewage Sanitation: An Integrated Study

Sewage needs to be treated properly before it is disposed. A few types of sanitation systems are combined sewer, effluent sewer, sanitary sewer, simplified sewerage, storm drain, and vacuum sewer. The various systems of maintaining sanitation are best understood in confluence with the major topics listed in the following section.

Sewage Sanitation

For safe disposal of the sewage generated from a locality efficient collection, conveyance, adequate treatment and proper disposal of treated sewage is necessary. To achieve this, following conditions should be satisfied:

1. Sewage should not pollute the drinking water source, either surface or groundwater, or water bodies that are used for bathing or recreational purposes.
2. The untreated sewage during conveyance should not be exposed so as to have access to human being or animals and should not give unsightly appearances or odour nuisance, and should not become a place for breeding flies.
3. It should not cause harm to public health and adversely affect the receiving environment.

The collection system is meant for collection of the sewage generated from individual houses and transporting it to a common point where it can be treated as per the needs before disposal. In olden days, waste generated from water closets was collected by conservancy methods and other liquid waste was transported through open drain to finally join natural drains. Since, the excreta was carried through carts, it was not hygienic method for transportation to the disposal point. Now, collection and conveyance of sewage is done in water carriage system, where it is transported in closed conduit using water as a medium.

Types of Sanitary Sewer Systems

Combined Sewer

A combined sewer is a sewage collection system of pipes and tunnels designed to also collect surface runoff. Combined sewers can cause serious water pollution problems during combined sewer overflow (CSO) events when wet weather flows exceed the sewage treatment plant capacity. This type of sewer design is no longer used in building new communities (because current design separates sanitary sewers from runoff), but many older cities continue to operate combined sewers.

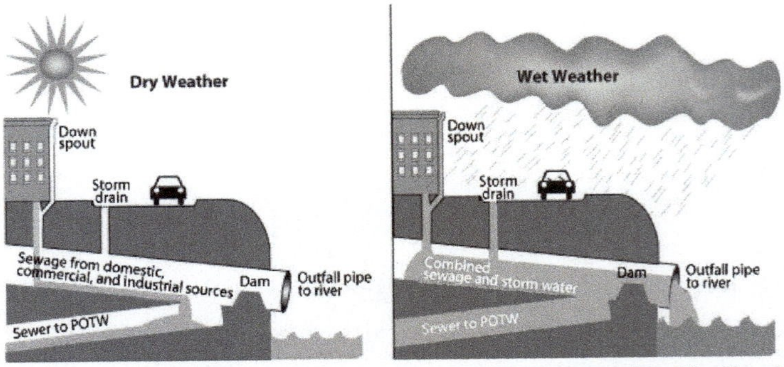

Combined Sewer System. During dry weather (and small storms), all flows are handled by the publicly owned treatment works (POTW). During large storms, the relief structure allows some of the combined stormwater and sewage to be discharged untreated to an adjacent water body.

Background

Recent archaeological discoveries have shown that some of the earliest sewer systems were developed 2500 BC in the ancient city of Harappa. The primitive sewers were carved in the ground alongside buildings. This discovery reveals the conceptual understanding of waste disposal by the early civilizations.

The earliest sewers were designed to carry street runoff away from inhabited areas and into surface waterways without treatment. Open sewers, consisting of gutters and urban streambeds, were common worldwide before the 20th century. In the majority of developed countries, large efforts were made during the late 19th and early 20th centuries to cover the formerly open sewers, converting them to closed systems with cast iron, steel, or concrete pipes, masonry, and concrete arches. Most sewage collection systems of the 19th and early to mid 20th century used single-pipe systems that collect both sewage and urban runoff from streets and roofs. This type of collection system is referred to as a combined sewer system. The rationale for combining the two was that it would be cheaper to build just a single system. Most cities at that time did not have sewage treatment plants, so there was no perceived public health advantage in constructing a separate "surface water sewerage" (the term used in the UK) or "storm sewer" (the term used in the U.S.) system.

When constructed, combined sewer systems were typically sized to carry three to 160 times the average dry weather sewage flows. It is generally infeasible to treat the volume of mixed sewage and surface runoff flowing in a combined sewer during peak runoff events caused by snowmelt or convective precipitation. As cities built sewage treatment plants, those plants were typically built to treat only the volume of sewage flowing during dry weather. Relief structures were installed in the collection system to bypass untreated sewage mixed with surface runoff during wet weather, protecting sewage treatment plants from damage caused if peak flows reached the headworks.

Combined Sewer Overflows (CSOs)

These relief structures, called storm-water regulators (in American English - or combined sewer overflows in British English) are constructed in combined sewer systems to divert flows in excess of the peak design flow of the sewage treatment plant. Combined sewers are built with control sections establishing stage-discharge or pressure differential-discharge relationships which may be either predicted or calibrated to divert flows in excess of sewage treatment plant capacity. A leaping weir may be used as a regulating device allowing typical dry-weather sewage flow rates to fall into an interceptor sewer to the sewage treatment plant, but causing a major portion of higher flow rates to leap over the interceptor into the diversion outfall. Alternatively, an orifice may be sized to accept the sewage treatment plant design capacity and cause excess flow to accumulate above the orifice until it overtops a side-overflow weir to the diversion outfall.

Photo of the interior of a combined sewer in Brighton, England.

CSO statistics may be confusing because the term may describe either the number of events or the number of relief structure locations at which such events may occur. A CSO event, as the term is used in American English, occurs when mixed sewage and stormwater are bypassed from a combined sewer system control section into a river, stream, lake, or ocean through a designed diversion outfall, but without treatment. Overflow frequency and duration varies both from system to system, and from outfall to outfall, within a single combined sewer system. Some CSO outfalls discharge infrequently, while others activate every time it rains.

CSOs should not be confused with sanitary sewer overflows. Sanitary sewer overflows are caused by sewer system obstructions, damage, or flows in excess of sewer capacity (rather than treatment plant capacity.) Sanitary sewer overflows may occur at any low spot in the sewer system rather than at the CSO relief structures. Absence of a diversion outfall often causes sanitary sewer overflows to flood residential structures and/or flow over traveled road surfaces before reaching natural drainage channels. Sanitary sewer overflows may cause greater health risks and environmental damage than CSOs if they occur during dry weather when there is no precipitation runoff to dilute and flush away sewage pollutants.

The storm water component contributes pollutants to CSO; but a major faction of pollution is the first foul flush of accumulated biofilm and sanitary solids scoured from the dry weather wetted perimeter of combined sewers during peak flow turbulence. Each storm is different in the quantity and type of pollutants it contributes. For example, storms that occur in late summer, when it has not rained for a while, have the most pollutants. Pollutants like oil, grease, fecal coliform from pet and wildlife waste, and pesticides get flushed into the sewer system. In cold weather areas, pollutants from cars, people and animals also accumulate on hard surfaces and grass during the winter and then are flushed into the sewer systems during heavy spring rains.

CSOs in the United States

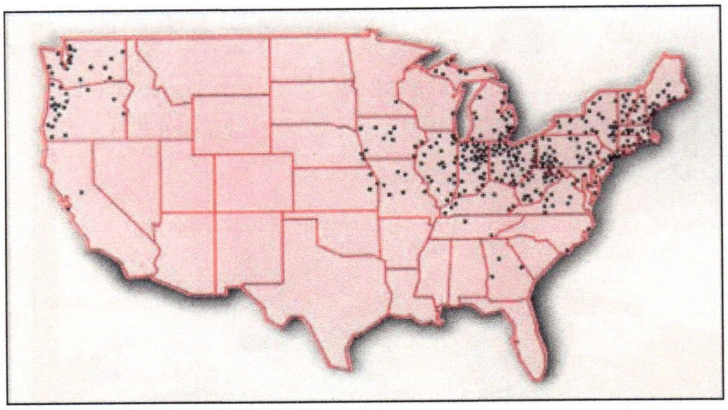

Most of the US combined sewer systems are in the Northeast and Great Lakes regions, and the Pacific Northwest.

About 860 communities in the US have combined sewer systems, serving about 40 million people. CSO discharges during heavy storms can cause serious water pollution problems in these communities. Pollutants from CSO discharges can include bacteria and other pathogens, toxic chemicals, and debris. These pollutants have also been linked with antimicrobial resistance, posing serious public health concerns. The U.S. Environmental Protection Agency (EPA) issued a policy in 1994 requiring municipalities to make improvements to reduce or eliminate CSO-related pollution problems. It is managed by the National Pollutant Discharge Elimination System (NPDES) permit program. The policy defined water quality parameters for the safety of an ecosystem; it allowed for action that are site specific to control CSOs in most practical way for community; it made sure the CSO control is not beyond a community's budget; and allowed water quality parameters to be flexible, based upon the site specific conditions. The CSO Control Policy required all states to have "nine minimum controls" in place by January 1, 1997, to decrease the effects of sewage overflow by making small improvements in existing processes. In 2000 Congress amended the Clean Water Act to require the municipalities to comply with the EPA policy.

Mitigation of CSOs

The United Kingdom Environment Agency identified unsatisfactory intermittent discharges and issued an Urban Wastewater Treatment Directive requiring action to limit pollution from combined sewer overflows. In 2009 the Canadian Council of Ministers of the Environment adopted a Canada-wide Strategy for the Management of Municipal Wastewater Effluent including national standards to (1) remove floating material from combined sewer overflows, (2) prevent combined sewer overflows during dry weather, and (3) prevent development or redevelopment from increasing frequency of combined sewer overflows.

Municipalities in the US have been undertaking projects to mitigate CSO since the 1990s. For example, prior to 1990, the quantity of untreated combined sewage discharged annually to lakes, rivers and streams in southeast Michigan was estimated at more than 30 billion US gallons (110,000,000 m^3) per year. In 2005, with nearly $1 billion of a planned $2.4 billion CSO investment put into operation, untreated discharges have been reduced by more than 20 billion US gallons (76,000,000 m^3) per year. This investment that has yielded an 85 percent reduction in CSO has included numerous sewer separation, CSO storage and treatment facilities and wastewater treatment plant improvements constructed by local and regional governments.

Many other areas in the US are undertaking similar projects. Cities like Pittsburgh, Seattle, Philadelphia, and New York are focusing on these projects partly because they are under federal consent decrees to solve their CSO issues. Both up-front penalties and stipulated penalties are utilized by EPA and state agencies to enforce CSO-mitigating initiatives and the efficiency of their schedules. Municipalities' sewage departments, engineering and design firms, and environmental organizations offer different approaches to potential solutions.

Sewer Separation

Some US cities have undertaken sewer separation projects — building a second piping system for all or part of the community. In many of these projects, cities have been able to separate only portions of their combined systems. High costs or physical limitations may preclude building a completely separate system. In 2011 Washington, D.C. separated its sewers in four small neighborhoods at a cost of $11 million. (The project cost also includes improvements to the drinking water piping system.)

CSO Storage

Another solution is to build a CSO storage facility, such as a tunnel that can store flow from many sewer connections. Because a tunnel can share capacity among several outfalls, it can reduce the total volume of storage that must be provided for a specific number of outfalls. Storage tunnels store combined sewage but do not treat it. When the storm is over, the flows are pumped out of the tunnel and sent to a wastewater treatment plant. One of the main concerns with CSO storage is the time it is stored before it is released. Without careful management of this time the water in the CSO storage facility runs the risk of going septic.

Washington, D.C. is building underground storage capacity as its primary strategy to address CSOs. As of 2011, the city is building a deep storage tunnel, adjacent to the Anacostia River, that will reduce overflows to the river by 98 percent, and 96 percent system-wide. (The city's overall "Clean Rivers" project, projected to cost $2.6 billion, includes other components, such as reducing stormwater flows.) The South Boston CSO Storage Tunnel is a similar project, completed in 2011.

Expanding Sewage Treatment Capacity

Some cities have expanded their basic sewage treatment capacity to handle some or all of the CSO volume. In 2002 litigation forced the city of Toledo, Ohio to double its treatment capacity and build a storage basin in order to eliminate most overflows. The city also agreed to study ways to reduce stormwater flows into the sewer system.

Retention Basins

Retention treatment basins or large concrete tanks that store and treat combined sewage are another solution. These underground structures can range in storage and treatment capacity from

2 million US gallons (7,600 m^3) to 120 million US gallons (450,000 m^3) of combined sewage. While each facility is unique, a typical facility operation is as follows. Flows from the overloaded sewers are pumped into a basin that is divided into compartments. The first flush compartment captures and stores flows with the highest level of pollutants from the first part of a storm. These pollutants include motor oil, sediment, road salt, and lawn chemicals (pesticides and fertilizers) that are picked up by the stormwater as it runs off roads and lawns. The flows from this compartment are stored and sent to the wastewater treatment plant when there is capacity in the interceptor sewer after the storm. The second compartment is a treatment or flow-through compartment. The flows are disinfected by injecting sodium hypochlorite, or bleach, as they enter this compartment. It then takes about 20-30 minutes for the flows to move to the end of the compartment. During this time, bacteria are killed and large solid materials settle out. At the end of the compartment, any remaining sanitary trash is skimmed off the top and the treated flows are discharged into the river or lake.

Screening and Disinfection Facilities

Screening and disinfection facilities treat CSO without ever storing it. Called "flow-through" facilities, they use fine screens to remove solids and sanitary trash from the combined sewage. Flows are injected with sodium hypochlorite for disinfection and mixed as they travel through a series of fine screens to remove debris. The fine screens have openings that range in size from 4 to 6 mm, or a little less than a quarter inch. The flow is sent through the facility at a rate that provides enough time for the sodium hypochlorite to kill bacteria. All of the materials removed by the screens are then sent to the sewage treatment plant through the interceptor sewer.

Reducing Stormwater Flows

Communities may implement low impact development techniques to reduce flows of stormwater into the collection system. This includes:

- constructing new and renovated streets, parking lots and sidewalks with interlocking stones, permeable paving and pervious concrete
- installing green roofs on buildings
- installing bioretention systems, also called rain gardens, in landscaped areas
- Rainwater harvesting equipment collects runoff from building roofs during wet weather for irrigating landscapes and gardens during dry weather
- Graywater collection and use on site reduces sewage discharges at all times

Gray vs. Green Infrastructure

CSO mitigating initiatives that are solely composed of sewer system reconstruction are referred to as gray infrastructure, while techniques like permeable pavement and rainwater harvesting are referred to as green infrastructure. Conflict often occurs between a municipality's sewage authority and its environmentally active organizations between gray and green infrastructural plans.

The 2004 EPA *Report to Congress* on CSO's provides a review of available technologies to mitigate CSO impacts.

New Approaches

Smart Infrastructure

Recent technological advances in sensing and control have enabled the implementation of Real Time Decision Support Systems (RT-DSS) for CSO mitigation. Through the use of internet of things technology and cloud computing, CSO events can now be mitigated by dynamically adjusting setpoints for movable gates, pump stations, and other actuated assets in sewers and storm water management systems. Similar technology, called adaptive traffic control is used to control the flow of vehicles through traffic lights. RT-DSS systems take advantage of storm temporal and spatial variability as well as varying concentration times due to diverse land uses across the sewershed to coordinate and optimize control assets. By maximizing storage and conveyance RT-DSS are able to minimize overflows using existing infrastructure. Successful implementations of RT-DSS have been carried out throughout the United States and Europe.

History

As a product of the Industrial Revolution, many cities in Europe and North America grew in the 19th century, frequently leading to crowding and increasing concerns about public health. As part of a trend of municipal sanitation programs in the late 19th and 20th centuries, many cities constructed extensive sewer systems to help control outbreaks of disease such as typhoid and cholera. Initially these systems discharged sewage directly to surface waters without treatment. As pollution of water bodies became a concern, cities added sewage treatment plants to their systems. Most cities in the Western world added more expensive systems for sewage treatment in the early 20th century.

As Britain was the first country to industrialize, it was also the first to experience the disastrous consequences of major urbanisation and was the first to construct a sewerage system as we know it today to mitigate the resultant unsanitary conditions. Joseph Bazalgette designed an extensive underground sewerage system that diverted waste to the Thames Estuary, downstream of the main centre of population. Six main interceptor sewers, totalling almost 135 miles (217 km) in length, were constructed. The intercepting sewers, constructed between 1859 and 1865, were fed by 450 miles (720 km) of main sewers that, in turn, conveyed the contents of some 13,000 miles (21,000 km) of smaller local sewers. With only minor modifications, Bazalgette's engineering achievement remains the basis for sewerage design up into the present day.

In France, the Paris cholera epidemic of 1832 sharpened the public awareness of the necessity for some sort of drainage system to deal with sewage and waster water in a better and healthier way since the Seine received up to 100,000 cubic meters of wastewater per day. Between 1865 and 1920 Eugene Belgrand led the development of a large scale system for water supply and wastewater management. By 1894 laws were passed which made drainage mandatory. The treatment of Paris sewage, though, was left to natural devices as 5,000 hectares of land were used to spread the waste out to be naturally purified.

Society and Culture

A combined sewer-pipe being laid by the city's sewerage company in Ghent, Belgium.

The image of the sewer recurs in European culture as they were often used as hiding places or routes of escape by the scorned or the hunted, including partisans and resistance fighters in World War II. Fighting erupted in the sewers during the Battle of Stalingrad. The only survivors from the Warsaw Uprising and Warsaw Ghetto made their final escape through city sewers. Some have commented that the engravings of imaginary prisons by Piranesi were inspired by the Cloaca Maxima, one of the world's earliest sewers.

Regulations

United Kingdom

There is in the UK a legal difference between a storm sewer and a surface water sewer. You do not have a right of connection to a storm sewer under section 106 of the water Industry Act. These are normally the pipe line that discharges to a watercourse, downstream of a combined sewer overflow. It takes the excess flow from a combined sewer. A surface water sewer conveys rainwater; legally you have a right of connection for your rainwater to this public sewer. A public storm water sewer can discharge to a public surface water, but not the other way around, without a legal change in sewer status by the water company.

In Fiction

The theme of traveling through, hiding, or even residing in combined sewers is a common plot device in media. Famous examples of sewer dwelling are the Teenage Mutant Ninja Turtles, Stephen King's *It*, *Les Miserables*, *The Third Man*, *Ladyhawke*,*Mimic*, *The Phantom of the Opera*, *Beauty and the Beast*, and *Jet Set Radio Future*. The Todd Strasser novel *Y2K-9: the Dog Who Saved the World* is centered around a dog thwarting terroristic threats to electronically sabotage American sewage treatment plants.

Sewer Alligators

A well-known urban legend, the sewer alligator, is that of giant alligators or crocodiles residing in combined sewers, especially of major metropolitan areas. Two public sculptures in New York depict an alligator dragging a hapless victim into a manhole.

Alligators have been known to get into combined storm sewers in the southeastern United States. Closed-circuit television by a sewer repair company captured an alligator in a combined storm sewer on tape.

Sanitary Sewer

PVC sanitary sewer installation. Sanitary sewers are sized to carry the amount of sewage generated by the collection area. Sanitary sewers are much smaller than combined sewers designed to also carry surface runoff.

A sanitary sewer or "foul sewer" is an underground carriage system specifically for transporting sewage from houses and commercial buildings through pipes to treatment or disposal. Sanitary sewers are part of an overall system called sewerage or sewage system.

Sewage may be treated to control water pollution before discharge to surface waters. Sanitary sewers serving industrial areas also carry industrial wastewater.

Separate sanitary sewer systems are designed to transport sewage alone. In municipalities served by sanitary sewers, separate storm drains may be constructed to convey surface runoff directly to surface waters. Sanitary sewers are distinguished from combined sewers, which combine sewage with stormwater runoff in the same pipe. Sanitary sewer systems are considered beneficial because they avoid combined sewer overflows.

Background

Sewage treatment is less effective when sanitary waste is diluted with stormwater, and combined sewer overflows occur when runoff from heavy rainfall or snowmelt exceeds hydraulic capacity of sewage treatment plants. To overcome these disadvantages, some cities built separate sanitary sewers to collect only municipal wastewater and exclude stormwater runoff collected in separate storm drains. The decision between a combined sewer system or two separate systems is mainly based on need for sewage treatment and cost of providing treatment during heavy rain events. Many cities with combined sewer systems built prior to sewage treatment have not replaced those sewer systems.

Types

Conventional Gravity Sewers

In the developed world, sewers are pipes from buildings to one or more levels of larger underground trunk mains, which transport the sewage to sewage treatment facilities. Vertical pipes, usually made of precast concrete, called manholes, connect the mains to the surface. Depending upon site application and use, these vertical pipes can be cylindrical, eccentric or concentric. The manholes are used for access to the sewer pipes for inspection and maintenance, and as a means to vent sewer gases. They also facilitate vertical and horizontal angles in otherwise straight pipelines.

Manhole access to sewer; person shows scale

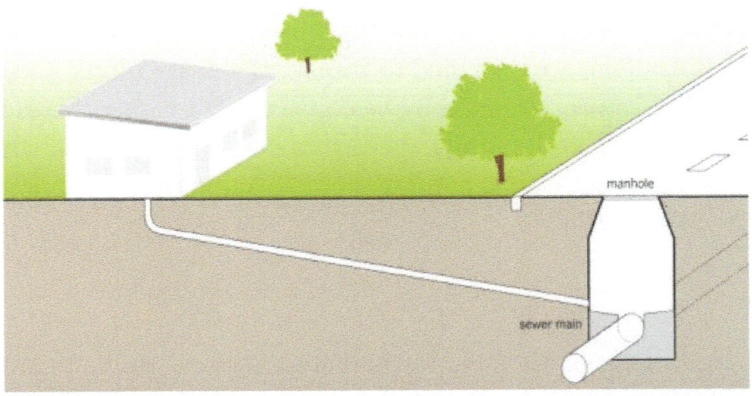

Schematic of a conventional sanitary sewer to convey blackwater and greywater from households to a centralized sewage treatment facility

Pipes conveying sewage from an individual building to a common gravity sewer line are called laterals. Branch sewers typically run under streets receiving laterals from buildings along that street and discharge by gravity into trunk sewers at manholes. Larger cities may have sewers called interceptors receiving flow from multiple trunk sewers.

A manhole cover for a sanitary sewer access point.

View looking down into an open manhole showing two converging sanitary sewer lines. The larger line enters from the right and changes direction within the manhole to exit from the top of the photo. A smaller line enters from the bottom of the photo under the access steps. The concrete floor of the manhole has channels to minimize accumulation of solids.

Interior photo of a large sanitary sewer from an access manhole.

Design and sizing of sanitary sewers considers the population to be served over the anticipated life of the sewer, per capita wastewater production, and flow peaking from timing of daily routines. Minimum sewer diameters are often specified to prevent blockage by solid materials flushed down toilets; and gradients may be selected to maintain flow velocities generating sufficient turbulence to minimize solids deposition within the sewer. Commercial and industrial flows are also considered, but diversion of surface runoff to storm drains eliminates wet weather flow peaks of inefficient combined sewers.

Force mains

Pumps may be necessary where gravity sewers serve areas at lower elevations than the sewage treatment plant, or distant areas at similar elevations. A lift station is a gravity sewer sump with a pump to lift accumulated sewage to a higher elevation. The pump may discharge to another gravity sewer at that location or may discharge through a pressurized force main to some distant location.

Effluent sewer

Effluent sewer systems, also called septic tank effluent drainage (STED) or solids-free sewer (SFS) systems, have septic tanks that collect sewage from residences and businesses, and the effluent that comes out of the tank is sent to either a centralized sewage treatment plant or a distributed treatment system for further treatment. Most of the solids are removed by the septic tanks, so the treatment plant can be much smaller than a typical plant. In addition, because of the vast reduction in solid waste, a pumping system can be used to move the wastewater rather than a gravity system. The pipes have small diameters (typically 1.5" to 4" (inches), or 4 to 10 cm). Because the wastestream is pressurized, they can be laid just below the ground surface along the land's contour.

Simplified Sewer

Simplified sanitary sewers consist of small-diameter pipes (typically 100 mm or about 4 inches), often laid at fairly flat gradients (1 in 200). The investment cost for simplified sanitary sewers can be about half the cost of conventional sewers. However, the requirements for operation and maintenance are usually higher. Simplified sewers are most common in Brazil and are also used in a number of other developing countries.

Vacuum Sewer

In low-lying communities, wastewater may be conveyed by vacuum sewer. Pipelines range in size from pipes of six inches (150 mm) in diameter to concrete-lined tunnels of up to thirty feet (10 m) in diameter. A low pressure system uses a small grinder pump located at each point of connection, typically a house or business. Vacuum sewer systems use differential atmospheric pressure to move the liquid to a central vacuum station.

Maintenance

Sanitary sewer overflow can occur due to blocked or broken sewer lines, infiltration of excessive stormwater or malfunction of pumps. In these cases untreated sewage is discharged from a sanitary sewer into the environment prior to reaching sewage treatment facilities. To avoid this, maintenance is required.

The maintenance requirements vary with the type of sanitary sewer. In general, all sewers deteriorate with age, but infiltration/inflow is a problem unique to sanitary sewers, since both combined sewers and storm drains are sized to carry these contributions. Holding infiltration to acceptable levels requires a higher standard of maintenance than necessary for structural integrity considerations of combined sewers. A comprehensive construction inspection program is required to prevent inappropriate connection of cellar, yard, and roof drains to sanitary sewers.

The probability of inappropriate connections is higher where combined sewers and sanitary sewers are found in close proximity, because construction personnel may not recognize the difference. Many older cities still use combined sewers while adjacent suburbs were built with separate sanitary sewers.

For decades, when sanitary sewer pipes cracked or experienced other damage, the only option was an expensive excavation, removal and replacement of the damaged pipe, typically requiring street repavement afterwards. In the mid-1950s a unit was invented where two units at each end with a special cement mixture in between was pulled from one manhole cover to the next, coating the pipe with the cement under high pressure which then cured rapidly, sealing all cracks and breaks in the pipe. Today, a similar method using epoxy resin is used by some municipalities to re-line aging or damaged pipes, effectively creating a "pipe in a pipe". These methods may be unsuitable for locations where the full diameter of the original pipe is required to carry expected flows, and may be an unwise investment if greater wastewater flows may be anticipated from population growth, increased water use, or new service connections within the expected service life of the repair.

Another popular method for replacing aged or damaged lines is called Pipe bursting, where a new pipe (typically PVC or ABS) is drawn thru the old pipe behind an "expander head" that breaks apart the old pipe as the new one is drawn through behind it.

These methods are most suitable for trunk sewers, since repair of lines with lateral connections is complicated by making provisions to receive lateral flows without accepting undesirable infiltration from inadequately sealed junctions.

History

Sanitary sewers evolved from combined sewers built where water was plentiful. Animal feces accumulated on city streets while animal-powered transport moved people and goods. Accumulations of animal feces encouraged dumping chamber pots into streets where night soil collection was impractical. Combined sewers were built to use surface runoff to flush waste off streets and move it underground to places distant from populated areas. Sewage treatment became necessary as population expanded, but treatment of diluted waste from combined sewers is more expensive than treating undiluted sewage.

Simplified Sewerage

Simplified sewerage, also called small-bore sewerage, is a sewer system that collects all household wastewater (blackwater and greywater) in small-diameter pipes laid at fairly flat gradients. Simplified sewers are laid in the front yard or under the pavement (sidewalk) or - if feasible - inside the back yard, rather than in the centre of the road as with conventional sewerage. It is suitable for existing unplanned low-income areas, as well as new housing estates with a regular layout. It allows for a more flexible design. With simplified sewerage it is crucial to have management arrangements in place to remove blockages, which are more frequent than with conventional sewers. It has been estimated that simplified sewerage reduces investment costs by *up to* 50% compared to conventional sewerage.

Simplified sewerage is sometimes also referred to as conventional sewerage with appropriate standards, implying that most conventional sewers are overdesigned.

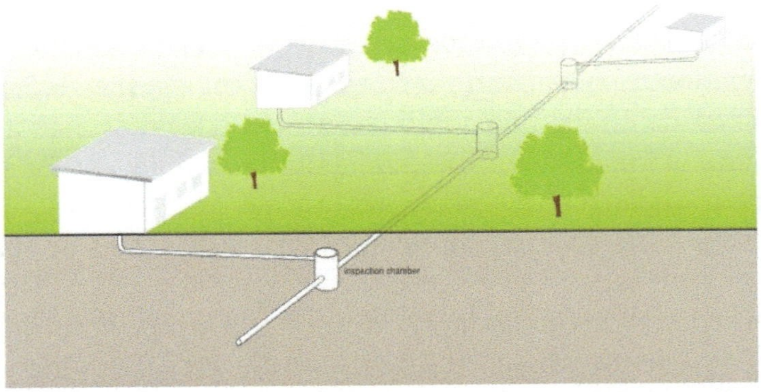

Schematic of a simplified sewer: Smaller diameter pipes are laid at a shallower depth and at a flatter gradient than for conventional sewers.

The concept of simplified sewerage emerged in parallel in Natal, Brazil and Karachi, Pakistan in the early 1980s without any interaction or communication. In both cases particular emphasis was given to community mobilization, an essential element for the success of simplified sewerage. In Latin America, and particularly in Brazil, simplified sewerage is also known as condominial sewerage, a term that undercores the importance of community participation in planning and maintenance at the level of a housing block (known as *condominio* in the Spanish and Portuguese use of the term).

Background

In developing countries, connection to sewer systems is often costly for poor households, despite typically low monthly sewer tariffs. This apparent paradox is explained by the high costs of in-plot and in-house sanitary installations that have to be paid entirely by the user, by sometimes high sewer connection fees levied by utilities, and by a lack of community consultation. As a result, in many cities in developing countries conventional sewers are laid at high costs under a street, while many users on that street do not connect to them. In Brazil, in some cities connection rates in the early 1990s were less than 40% of the intended beneficiary population.

Application

Simplified sewerage is most widely used in Brazil. It is estimated that in Brazil some 5 million people in over 200 towns and cities are served with simplified sewerage - or condominial sewerage. This corresponds to about 3% of the population of Brazil and about 6% of the population connected to sewers. They serve poor and rich alike.

Simplified sewerage has also been used in

- Bolivia, beginning with a pilot project in El Alto;
- Honduras, primarily in marginal areas of Tegucigalpa where simplified sewerage has been introduced in 20 communities with 24,000 inhabitants;

- Peru, primarily in marginal areas of Lima;
- in South Africa, where pilot projects were carried out in Johannesburg and Durban;
- in Sri Lanka, where the National Housing Development Authority implemented over 20 schemes in the 1980s and 90s.

In Pakistan, beginning with the Orangi Pilot Project in Karachi, a variation of simplified sewerage using larger diameter pipes has been used.

Community Participation

Community participation in the planning of any sewer system is a fundamental requirement to achieve higher household connection rates and to increase the likelihood of proper maintenance of in-block sewers. In addition, it can motivate users to assume parts of the costs of the sewer system that they are able to assume, such as contribution of labor for construction and/ or maintenance.

Typically, in the planning process for a simplified sewerage system, meetings are carried out at the housing block (*condominio*) level for information, discussions and clarifications required for a joint group decision on network design, community contributions during construction and maintenance responsibilities. Users might finance and implement in-house sanitary installations and household connections and would agree on a suitable type of condominial branch. They are asked to comply with agreements established for construction and operation of the condominial branch, as well as payment of tariffs. In turn, the service provider agrees to fulfill his responsibilities as established in the "Terms of Connection " between the parties. The community participation process also provides a good opportunity for complementary actions like hygiene promotion, which can have a significant impact on public health at a relatively limited cost.

Design and Construction

Simplified sewers are usually laid in the front yard or under the pavement (sidewalk). In some rare cases it is possible to lay them in the back yard. Sidewalk branches are usually preferred in regular urbanizations, while the front and back yard branches are particularly suited to neighborhoods with challenging topography or urbanization patterns. However, in some cases neither of these options is possible. For example, in South Asia, in many cities there is no sidewalk or front yard, so pipes have to be laid in the middle of the street as with conventional sewers.

In Latin America typical simplified sewer diameters are 100 mm, laid at a gradient of 1 in 200 (0.5 percent). Such a sewer will serve around 200 households of 5 people with a wastewater flow of 80 litres per person per day. In Pakistan, however, there are no rigorous standards for sewer diameters. In a small pilot as part of the Orangi Pilot Project pipes with a diameter of 150mm were used.

Laying small diameter pipes at fairly flat gradients requires careful construction techniques. Plastic pipes are best used as they are more easily jointed correctly. This reduces wastewater leakage from the sewer and groundwater infiltration into it. With simplified sewerage there is no need to have the large expensive manholes of the type used for conventional sewerage — simple brick or plastic junction chambers are used instead.

Construction can be carried out by contractors or by trained and properly supervised community members. Training and proper supervision are actually needed in both cases, since contractors in many cities are not familiar with simplified sewerage.

Investment Cost Comparison

The cost of sewerage - conventional or simplified - are always site-specific, and estimates are subject to controversies. Construction costs of simplified sewerage are *up to* half the costs of conventional sewerage. Investment cost savings come from various design features that may or may not be present in a particular simplified sewerage system. Cost-saving features of any simplified sewerage system are a smaller diameter of pipes, smaller and shallower trenches and simplified manholes. The two latter features are estimated to account for most of the cost savings. Other features that could further reduce costs may only be present in some systems, such as:

- shorter networks;
- avoidance of the need to damage pavements and sidewalks (if they already exist and if pipes are laid in front or back yards);
- decentralized, small-scale wastewater treatment, and consequently elimination of main collectors and sewage pumping stations.

An element that may slightly increase costs compared to conventional sewerage is the introduction of grease traps or of interceptor chambers for the settlement of sludge. The latter are more common in South Asia and are not used in the condominial model. A 2006 study of four countries showed cost savings of 31-57% from the use of simplified sewerage compared to conventional sewerage with unit costs varying from US$119 per connection in a neighborhood in Bolivia and to US$759 per connection in a small town in Paraguay. A detailed estimate gives the costs of simplified sewerage in Lima as at least US$700 per household (US$120–140 per person), including in-house sanitary facilities (US$100 per household) and including design, supervision and social intermediation costs (US$126 per household, which are common costs shared with water infrastructure), but excluding taxes.

In general, at higher population densities sewer systems are cheaper than on-site sanitation (such as septic tanks). The switching value at which sewerage becomes less costly is largely determined by the type of sewerage, conventional or simplified. A 1983 study in Natal showed that the investment costs for simplified sewerage were lower than for on-site systems at the quite low population density of about 160 people per hectare. Conventional sewerage, however, was cheaper only at densities above 400 people per hectare.

Operation and Maintenance

Good operation and maintenance (O&M) is essential for the long-term sustainability of any sewerage system, but particularly for simplified sewerage, since the low diameter of pipes and low gradients make the system more vulnerable to clogging.

The original concept of householders being responsible for O&M of in-block condominial sewers has not worked well in the long term. A study of simplified sewerage systems in Brazil has shown that effective maintenance of sewers by utilities has often been the result of community pressure by neighborhood associations. Without such pressure maintenance by utilities has often been inadequate, and community maintenance has not come about either.

Therefore, alternative management systems had to be developed and a few examples are provided below:

- In rural Ceará a villager is employed by the Residents' Association to maintain the sewers and the wastewater treatment plant (typically, a single facultative waste stabilization pond). He is also responsible for the water supply.
- In parts of Recife in northeast Brazil the state water and sewerage company employs local contracting firms for O&M. Usually this is done by a small team comprising a technician engineer and two laborers who work in a low-income area served by simplified sewerage and to whom residents report any problems.
- In Brasília the water and sewerage company, which has over 1,200 km of condominial sewers, uses van-mounted water jet units to clear any blockages.

Concerning maintenance costs, available information indicates similar costs and requirements for the simplified and the conventional system under the same conditions. Simplified systems typically require more interventions, but the cost per intervention is lower. Comparative analytical studies are not yet available, however.

Constraints for Application

According to Jose Carlos Melo, who is considered to be the "father" of condominial sewers in Brazil, some important constraints for the application of simplified sewerage are:

- Lack of information on fundamentals and techniques of the approach or lack of experience in its application,
- Resistance to change: Institutional, technical and operational changes required by the service provider for implementing the condominial approach usually provoke resistance and can hinder the application.
- Normative and legal restrictions: Existing conservative design and construction standards linked to conventional systems can be an essential constraint in the introduction and dissemination of the systems.

Over the last years, countries like Bolivia and Peru reviewed and modernized technical standards according to methods and criteria established and accepted in Brazil in the 1980s, thus overcoming the latter constraint.

Advantages and Disadvantages of Combined System

Advantages

- In an area where rainfall is spread throughout a year, there is no need of flushing of sewers, as self cleansing velocity will be developed due to more quantity because of addition of storm water.

- Only one set of pipe will be required for house plumbing.
- In congested areas it is easy to lay only one pipe rather than two pipes as required in other systems.

Disadvantages

- Not suitable for the area with small period of rainfall in a year, because dry weather flow will be small due to which self cleansing velocity may not develop in sewers, resulting in silting.
- Large flow is required to be treated at sewage treatment plant before disposal, hence resulting in higher capital and operating cost of the treatment plant.
- When pumping is required this system is uneconomical.
- During rains overflowing of sewers will spoil public hygiene.

Advantages and Disadvantages of Separate System Advantages

Advantages

- As sewage flows in separate pipe, hence the quantity to be treated at sewage treatment plant is small, resulting in economy of treatment.
- This system may be less costly as only sanitary sewage is transported in closed conduit and storm water can be collected and conveyed through open drains.
- When pumping is required during disposal, this system is economical due to less flow.

Disadvantages

- Self cleansing velocity may not developed at certain locations in sewers and hence flushing of sewers may be required.
- This system requires laying two sets of pipe, which may be difficult in congested area.
- This system will require maintenance of two sets of pipelines and hence maintenance cost is more.

Advantages and Disadvantages of Partially Separate System

Advantages

- Economical and reasonable size sewers are required.
- Work of house plumbing is reduced as rain water from roofs, sullage from bathrooms and kitchen, etc. are combined with discharge from water closets.
- Flushing of sewers may not be required as small portion of storm water is allowed to enter in sanitary sewage.

Disadvantages

- Increased cost of pumping as compared to separate system at treatment plants and intermediate pumping station wherever required.
- In dry weather self-cleansing velocity may not develop in the sewers.

Considerations for the Type of System

Following points are considered before finalizing the type of collection system.

- The separate system requires laying of two sets of conduits whereas in combined system only one bigger size conduit is required.
- Laying of two separate conduits may be difficult in the congested streets.
- In combined system sewers are liable for silting during non-monsoon season, hence they are required to be laid at steeper gradients. Steeper gradients for the sewers may require more number of pumping stations, particularly for flat terrain, which may make the system costly.
- Large quantity of wastewater is required to be treated before discharge in case of combined system. Hence, large capacity treatment plant is required.
- In separate system, only sewage is treated before it is discharged into natural water body or used for irrigation. No treatment is generally given to the rainwater collected before it is discharge in to natural water body.
- In case of separate system pumping is only required for sewage. Pumping can be avoided for storm water lines, as they are not very deep and normally laid along the natural slopes.

  In combined system large capacity pumping station is required to safely handle the flow that is likely to be generated during highest design storm considered.
- Based on site conditions the economy of the system needs to be evaluated and selection is made accordingly.

Patterns of Collection System

The network of sewers consists of house sewers discharging the sewage to laterals. The lateral discharges the sewage into branch sewers or sub-mains and sub-mains discharge it into main sewer or trunk sewer. The trunk sewer carries sewage to the common point where adequate treatment is given to the sewage and then it is discharged. The patterns of collection system depend upon:

1. The topographical and hydrological features of the area.
2. The location and methods of treatment and disposal works.

3. The type of sewerage system employed, and
4. Extent of area to be served.

Following patterns can be adopted for collection systems as per the suitability (Birdie, 1990).

Perpendicular Pattern

- The shortest possible path is maintained for the rains carrying storm water and sewage.
- It is suitable for separate system and partially separate system for storm water drains.
- This pattern is not suitable for combined system, because treatment plant is required to be installed at many places; otherwise it will pollute the water body where the sewage is discharged.

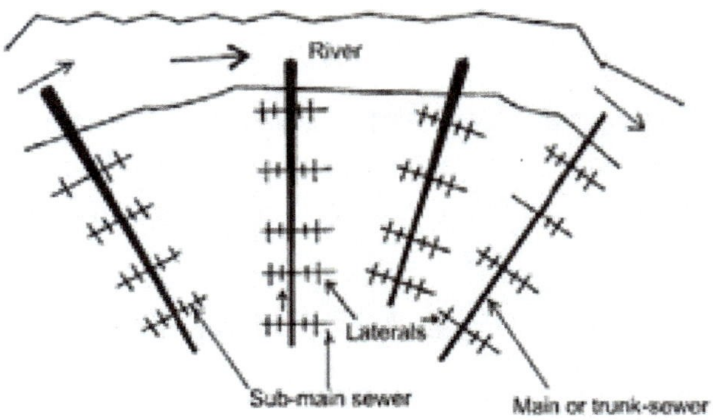

Perpendicular pattern of collection system

Interceptor Pattern

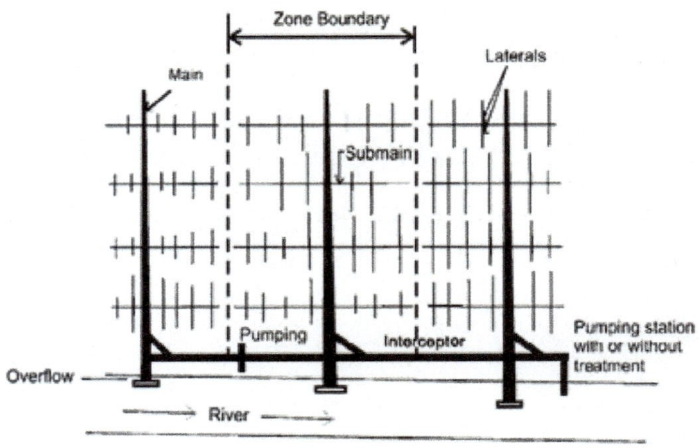

Interceptor pattern of collection system

- Sewers are intercepted with large size sewers.
- Interceptor carries sewage to a common point, where it can be disposed off with or without treatment.
- Overflows should be provided to handle very large flow.

Radial Pattern

- It is suitable for land disposal.
- In this pattern sewers are laid radialy outwards from the centre, hence this pattern is called as radial pattern.
- The drawback in this pattern is more number of disposal works are required.

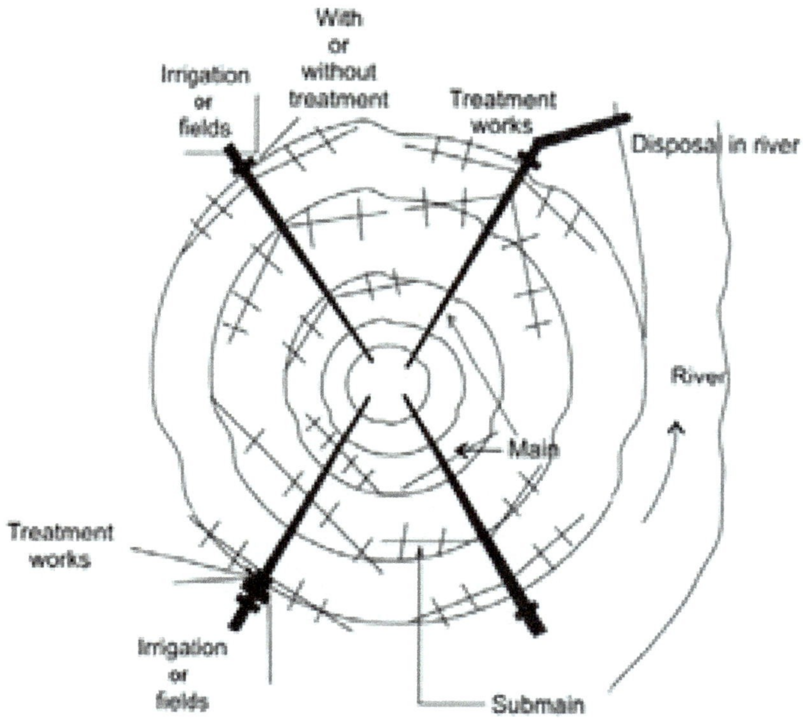

Radial pattern of collection system

Fan Pattern

- This pattern is suitable for a city situated at one side of the natural water body, such as river.
- The entire sewage flows to a common point where one treatment plant is located.
- In this number of converging main sewers and sub-mains are used forming a fan shape.
- Single treatment plant is required in this pattern.
- The drawback in this pattern is that larger diameter sewer is required near to the treatment plant as entire sewage is collected at a common point.
- In addition, with new development of the city the load on existing treatment plant increases.

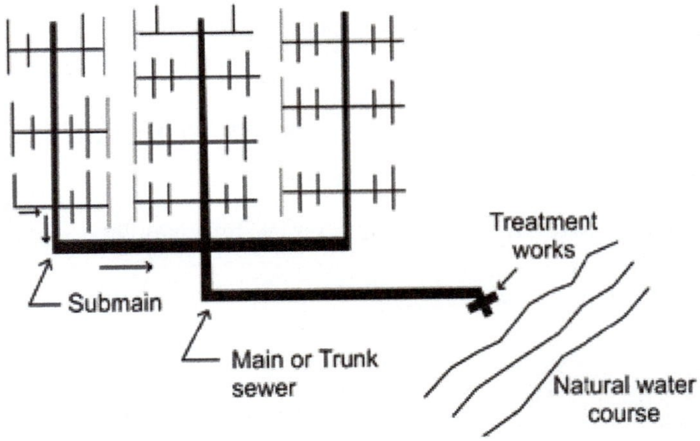

Fan pattern of collection system

Zone Pattern

- More numbers of interceptors are provided in this pattern.
- This pattern is suitable for sloping area than flat areas.

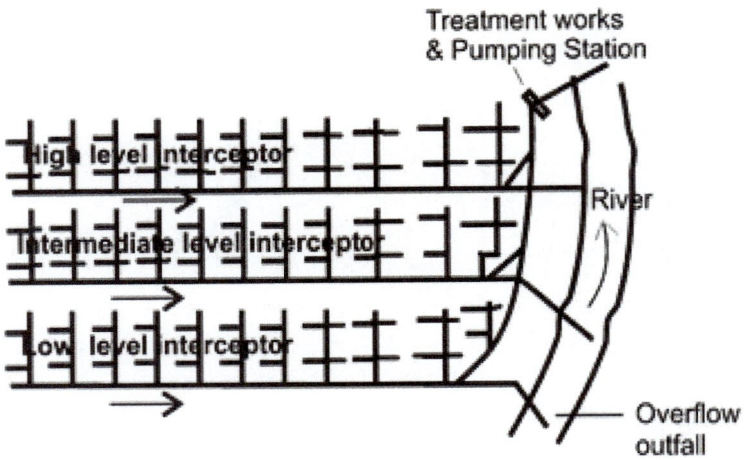

Zone pattern of collection system

Quantity Estimation of Sewage

The sewage collected from the municipal area consists of wastewater generated from the residences, commercial centers, recreational activities, institutions and industrial wastewaters discharge into sewer network from the permissible industries located within the city limits. Before designing the sewer, it is necessary to know the discharge i.e., quantity of sewage, which will flow in it after completion of the project.

Accurate estimation of sewage discharge is necessary for hydraulic design of the sewers. Far lower estimation than reality will soon lead to inadequate sewer size after commissioning of the scheme or the sewers may not remain adequate for the entire design period. Similarly, very high discharge estimated will lead to larger sewer size affecting economy of the sewerage scheme, and the lower discharge actually flowing in the sewer may not meet the criteria of the self cleansing velocity and hence leading to deposition in the sewers. Actual measurement of the discharge is not possible if the sewers do not exist; and where the capacity of the existing sewers is inadequate and need to be increased, still actual present discharge measurement may not be accurate due to unaccounted overflow and leakages that might be occurring in the existing system. Since sewers are design to serve for some more future years, engineering skills have to be used to accurately estimate the sewage discharge.

Sources of Sanitary Sewage

1. Water supplied by water authority for domestic usage, after desired use it is discharged in to sewers as sewage.
2. Water supplied to the various industries for various industrial processes by local authority. Some quantity of this water after use in different industrial applications is discharged as wastewater.
3. The water supplied to the various public places such as, schools, cinema theaters, hotels, hospitals, and commercial complexes. Part of this water after desired use joins the sewers as wastewater.
4. Water drawn from wells by individuals to fulfill domestic demand. After uses this water is discharged in to sewers.
5. The water drawn for various purposes by industries, from individual water sources such as, wells, tube wells, lake, river, etc. Fraction of this water is converted into wastewater in different industrial processes or used for public utilities within the industry generating wastewater. This is discharged in to sewers.
6. Infiltration of groundwater into sewers through leaky joints.
7. Entrance of rainwater in sewers during rainy season through faulty joints or cracks in sewers.

Dry Weather Flow

Dry weather flow is the flow that occurs in sewers in separate sewerage system or the flow that occurs during dry seasons in combined system. This flow indicates the flow of sanitary sewage. This depends upon the rate of water supply, type of area served, economic conditions of the people, weather conditions and infiltration of groundwater in the sewers, if sewers are laid below groundwater table.

Evaluation of Sewage Discharge

Correct estimation of sewage discharge is necessary; otherwise sewers may prove inadequate resulting in overflow or may prove too large in diameter, which may make the system uneconomical and hydraulically inefficient. Hence, before designing the sewerage system it is important to know the discharge / quantity of the sewage, which will flow in it after completion of the project and at the end of design period.

Apart from accounted water supplied by water authority that will be converted to wastewater, following quantities are considered while estimating the sewage quantity:

Addition Due to Unaccounted Private Water Supplies

People using water supply from private wells, tube wells, etc. contribute to the wastewater generation more than the water supplied by municipal authority. Similarly, certain industries utilize their own source of water. Part of this water, after desired uses, is converted into wastewater and ultimately discharged into sewers. This quantity can be estimated by actual field observations.

Addition Due to Infiltration

This is additional quantity due to groundwater seepage in to sewers through faulty joints or cracks formed in the pipes. The quantity of the water depends upon the height of the water table above the sewer invert level. If water table is well below the sewer invert level, the infiltration can occur only after rain when water is moving down through soil. Quantity of the water entering in sewers depends upon the permeability of the ground soil and it is very difficult to estimate. While estimating the design discharge, following suggested discharge can be considered.

Table Suggested estimates for groundwater infiltration for sewers laid below groundwater table (CPHEEO Manual, 1993)

Unit	Minimum	Maximum
L/ha.d	5000	50000
L/km.d	500	5000
L per day per manhole	250	500

Storm water drainage may also infiltrate into sewers. This inflow is difficult to calculate. Generally, no extra provision is made for this quantity. This extra quantity can be taken care of by extra empty space left at the top in the sewers, which are designed for running ¾ full at maximum design discharge.

Subtraction Due to Water Losses

The water loss, through leakage in water distribution system and house connections, does not reach consumers and hence, not appear as sewage.

Subtraction Due to Water not Entering the Sewerage System

Certain amount of water is used for such purposes, which may not generate sewage, e.g. boiler feed water, water sprinkled over the roads, streets, lawns, and gardens, water consumed in industrial product, water used in air coolers, etc.

Net quantity of sewage: The net quantity of sewage production can be estimated by considering the addition and subtraction as discussed above over the accounted quantity of water supplied by water authority as below:

$$\text{Net quantity of sewage} = \begin{matrix}\text{Accounted}\\ \text{quantity of}\\ \text{water supplied,}\\ \text{from the}\\ \text{waterworks}\end{matrix} \begin{matrix}\text{Addition due to}\\ \text{unaccounted}\\ \text{private water}\\ \text{supplies}\end{matrix} + \begin{matrix}\text{Additiondue}\\ \text{to infiltration}\end{matrix} - \begin{matrix}\text{Subtraction}\\ \text{due to water}\\ \text{not entering}\\ \text{the sewerage}\\ \text{system}\end{matrix}$$

Generally, 75 to 80% of accounted water supplied is considered as quantity of sewage produced.

Variation in Sewage Flow

Variation occurs in the flow of sewage over annual average daily flow. Fluctuation in flow occurs from hour to hour and from season to season. The typical hourly variation in the sewage flow is shown in the figure below. If the flow is gauged near its origin, the peak flow will be quite pronounced. The peak will defer if the sewage has to travel long distance. This is because of the time required in collecting sufficient quantity of sewage required to fill the sewers and time required in travelling. As sewage flow in sewer lines, more and more sewage is mixed in it due to continuous increase in the area being served by the sewer line. This leads to reduction in the fluctuations in the sewage flow and the lag period goes on increasing. The magnitude of variation in the sewage quantity varies from place to place and it is very difficult to predict. For smaller township this variation will be more pronounced due to lower length and travel time before sewage reach to the main sewer and for large cities this variation will be less.

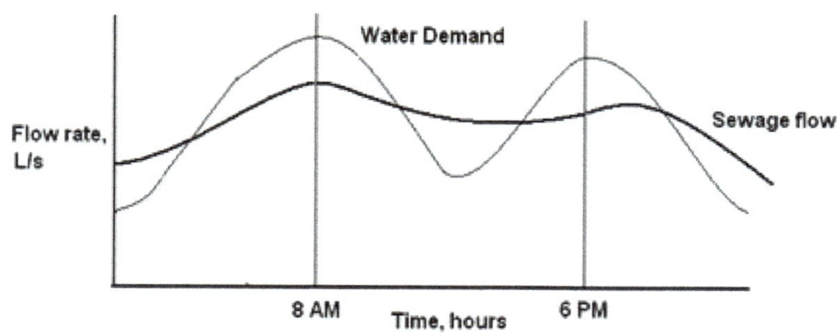

Typical hourly variations in sewage flow

For estimating design discharge following relation can be considered:

Maximum daily flow = Two times the annual average daily flow (representing seasonal variations)

Maximum hourly flow= 1.5 times the maximum daily flow (accounting hourly variations)

= Three times the annual average daily flow

As the tributary area increases, peak hourly flow will decrease. For smaller population served (less than 50000) the peak factor can be 2.5, and as the population served increases its value reduces. For large cities it can be considered about 1.5 to 2.0. Therefore, for outfall sewer the peak flow can be considered as 1.5 times the annual average daily flow. Even for design of the treatment facility, the peak factor is considered as 1.5 times the annual average daily flow.

The minimum flow passing through sewers is also important to develop self cleansing velocity to avoid silting in sewers. This flow will generate in the sewers during late night hours. The effect of this flow is more pronounced on lateral sewers than the main sewers. Sewers must be checked for minimum velocity as follows:

Minimum daily flow = 2/3 Annual average daily flow

Minimum hourly flow = ½ minimum daily flow

= 1/3 Annual average daily flow

The overall variation between the maximum and minimum flow is more in the laterals and less in the main or trunk sewers. This ratio may be more than 6 for laterals and about 2 to 3 in case of main sewers.

Design Period

The future period for which the provision is made in designing the capacities of the various components of the sewerage scheme is known as the design period. The design period depends upon the following:

- Ease and difficulty in expansion,
- Amount and availability of investment,
- Anticipated rate of population growth, including shifts in communities, industries and commercial investments,
- Hydraulic constraints of the systems designed, and
- Life of the material and equipment.

Following design period can be considered for different components of sewerage scheme.

1. Laterals less than 15 cm diameter : Full development
2. Trunk or main sewers : 40 to 50 years
3. Treatment Units : 15 to 20 years
4. Pumping plant : 5 to 10 years

Design Discharge of Sanitary Sewage

The total quantity of sewage generated per day is estimated as product of forecasted population at the end of design period considering per capita sewage generation and appropriate peak factor. The per capita sewage generation can be considered as 75 to 80% of the per capita water supplied per day. The increase in population also result in increase in per capita water demand and hence, per capita production of sewage. This increase in water demand occurs due to increase in living standards, betterment in economical condition, changes in habit of people, and enhanced demand for public utilities.

Quantity Estimation of Storm Water

The surface run-off resulting after precipitation contributes to the stormwater. The quantity of stormwater reaching to the sewers or drains is very large as compared with sanitary sewage. The factors affecting the quantity of stormwater flow are as below:

i. Area of the catchment

ii. Slope and shape of the catchment area

iii. Porosity of the soil

iv. Obstruction in the flow of water as trees, fields, gardens, etc.

v. Initial state of catchment area with respect to wetness.

vi. Intensity and duration of rainfall

vii. Atmospheric temperature and humidity

viii. Number and size of ditches present in the area

Measurement of Rainfall

The rainfall intensity could be measured by using rain gauges and recording the amount of rain falling in unit time. The rainfall intensity is usually expressed as mm/hour or cm/hour. The rain gauges used can be manual recording type or automatic recording rain gauges.

Methods for Estimation of Quantity of Storm Water

1. Rational Method
2. Empirical formulae method

In both the above methods, the quantity of storm water is considered as function of intensity of rainfall, coefficient of runoff and area of catchment.

Time of Concentration: The period after which the entire catchment area will start contributing to the runoff is called as the time of concentration.

- The rainfall with duration lesser than the time of concentration will not produce maximum discharge.
- The runoff may not be maximum even when the duration of the rain is more than the time of concentration. This is because in such cases the intensity of rain reduces with the increase in its duration.

- The runoff will be maximum when the duration of rainfall is equal to the time of concentration and is called as critical rainfall duration. The time of concentration is equal to sum of inlet time and time of travel.

Time of concentration = Inlet time + time of travel

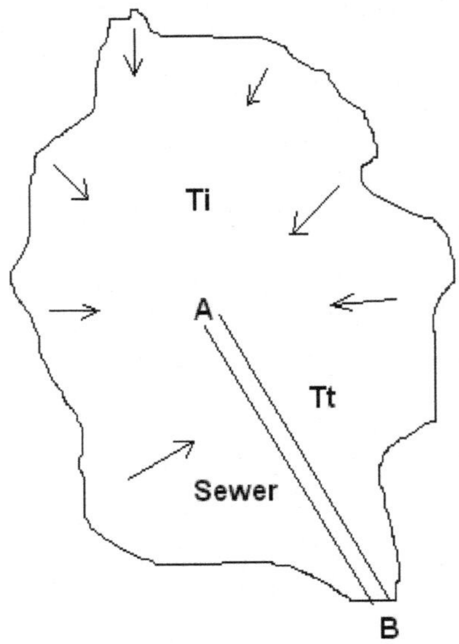

Runoff from a given catchment

Inlet Time: The time required for the rain in falling on the most remote point of the tributary area to flow across the ground surface along the natural drains or gutters up to inlet of sewer is called inlet time. The inlet time 'Ti' can be estimated using relationships similar to following. These coefficients will have different values for different catchments.

$$Ti = [0.885\ L^3/H]^{0.385} \quad (1)$$

Where,

Ti = Time of inlet, minute

L = Length of overland flow in Kilometer from critical point to mouth of drain

H = Total fall of level from the critical point to mouth of drain, meter

Time of Travel: The time required by the water to flow in the drain channel from the mouth to the point under consideration or the point of concentration is called as time of travel.

Time of Travel (Tt) = Length of drain/ velocity in drain (2)

Runoff Coefficient: The total precipitation falling on any area is dispersed as percolation, evaporation, storage in ponds or reservoir and surface runoff. The runoff coefficient can be defined as a fraction, which is multiplied with the quantity of total rainfall to determine the quantity of rain water, which will reach the sewers. The runoff coefficient depends upon the porosity of soil cover, wetness and ground cover. The overall runoff coefficient for the catchment area can be worked out as follows:

Overall runoff coefficient, $C = [A_1.C1 + A_2.C_2 + ... + A_n.C_n] / [A_1 + A_2 + ... + A_n]$ (3)

Where, A_1, A_2, ...An are types of area with C_1, C_2, ...C_n as their coefficient of runoff, respectively.

The typical runoff coefficient for the different ground cover is provided in the table below. The Table Runoff coefficient for different type of cover in catchment

Type of Cover	Coefficient of runoff
Business areas	0.70 – 0.90
Apartment areas	0.50 – 0.70
Single family area	0.30 – 0.50
Parks, Playgrounds, Lawns	0.10 – 0.25
Paved Streets	0.80 –0.90
Water tight roofs	0.70 – 0.95

Rational Method

Storm water quantity can be estimated by rational method as below:

Storm water quantity, $Q = C.I.A / 360$ (4)

Where,

Q = Quantity of storm water, m^3/sec

C = Coefficient of runoff

I = intensity of rainfall (mm/hour) for the duration equal to time of concentration, and

A = Drainage area in hectares

OR

$Q = 0.278\, C.I.A$ (5)

Where, Q is m^3/sec; I is mm/hour, and A is area in square kilometer

Empirical Formulae

Empirical formulae are used for determination of runoff from very large area. Various empirical relationships are developed based on the past observations on specific site conditions suiting a particular region. These empirical formulae can be used for prediction of storm water runoff for that particular catchment.

A] Burkli – Zeiglar formula

$$Q = \frac{C.I.A}{141.58} \sqrt[4]{S/A}$$

B] Mc Math formula (used in USA)

$$Q = \frac{C.I.A}{148.35} \sqrt[5]{S/A} \quad (6)$$

C] Fuller's formula

$$Q = \frac{C.M^{0.8}}{13.23} \quad (7)$$

(Where, S- Slope of the area in meter per thousand meter, M- drainage area in sq. km., A – drainage area in hectare)

Empirical Formulae for Rainfall Intensities

These relationships between rainfall intensity and duration are developed based on long term experience in field. Under Indian conditions, intensity of rainfall in design is usually in the range 12 mm/h to 20 mm/h. In general, the empirical relationship has the following forms:

$$I = a/(t + b) \quad \text{OR} \quad I = b / t^n \quad (8)$$

Where, a, b, and n are constants.

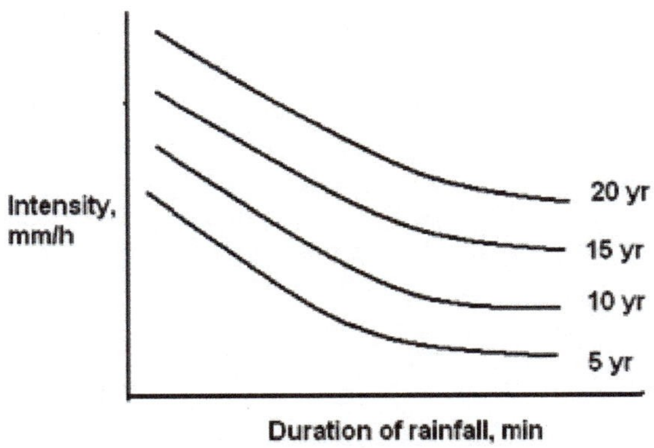

Relationship of rainfall duration and intensity

British Ministry of Health formula

$I = 760 / (t + 10)$ (for storm duration of 5 to 20 minutes) (9)

$I = 1020 / (t + 10)$ (for storm duration of 20 to 100 minutes) (10)

Where, I is intensity of rainfall, mm/h and t is duration of storm, minutes.

References

- Dhiman, Gaurav (2016). "Using Multiple Antibiotic Resistance Profiles of Coliforms as a Tool to Investigate Combined Sewer Overflow Contamination". Journal of Environmental Health
- "Going Against the Flow: Green Tech, Sensors and Industrial Internet Make Sewer Systems Smart". Txchnologist. General Electric. 2013. Retrieved October 16, 2015

- Benidickson, Jamie (2011). The Culture of Flushing: A Social and Legal History of Sewage. UBC Press. ISBN 9780774841382. Retrieved 2013-02-07
- Steven J. Burian, Stephan J. Nix, Robert E. Pitt, and S. Rocky Durrans (2000). "Urban Wastewater Management in the United States: Past, Present, and Future." Journal of Urban Technology, Vol. 7, No. 3, pp. 33-62. doi:10.1080/713684134
- L. Montestruque, M. Lemmon (2015). "Globally Coordinated Distributed Storm Water Management System." 1st International Workshop on Cyber Physical Systems for Smart Water Networks, doi:10.1145/2738935.2738948.
- Tilley, E., Ulrich, L., Lüthi, C., Reymond, Ph., Zurbrügg, C. (2014) Compendium of Sanitation Systems and Technologies - (2nd Revised Edition). Swiss Federal Institute of Aquatic Science and Technology (Eawag), Duebendorf, Switzerland. ISBN 978-3-906484-57-0
- Fan, Chi-Yuan; Field, Richard; Lai, Fu-hsiung. "Sewer-Sediment Control: Overview of an EPA Wet-Weather Flow Research Program" (PDF). University of California Los Angeles. United States Environmental Protection Agency. Retrieved 12 March 2016

4

A Comprehensive Study of Sewers and Manhole

Sewers are normally steeply inclined and are large in size to help free flow of huge amounts of sewage. A sewage system requires a proper design implementation to maximize its load and efficiency. It should be two to three metres below ground level and direct the sewage towards a lower area, and towards the treatment plant. The topics discussed in the chapter are of great importance to broaden the existing knowledge on wastewater management.

Hydraulic Design of Sewers and Storm Water Drains

Generally, sewers are laid at steeper gradients falling towards the outfall point with circular pipe cross section. Storm water drains are separately constructed as surface drains at suitable gradient, either rectangular or trapezoidal section. Sewers are designed to carry the maximum quantity of sanitary sewage likely to be produced from the area contributing to the particular sewer. Storm water drains are designed to carry the maximum storm runoff that is likely to be produced by the contributing catchment area from a rain of design frequency and of duration equal to the time of concentration.

Requirements of Design and Planning of Sewerage System

The sewerage scheme is designed to remove entire sewage effectively and efficiently from the houses to the point of treatment and disposal. Following aspects should be considered while designing the system.

- The sewers provided should be adequate in size to avoid overflow and possible health hazards.
- For evaluating proper diameter of the sewer, correct estimation of sewage discharge is necessary.
- The flow velocity inside the sewer should neither be so large as to require heavy excavation and high lift pumping, nor should be so small causing deposition of the solid in the sewers.
- The sewers should be laid at least 2 to 3 m deep to carry sewage from basement.
- The sewage in sewer should flow under gravity with 0.5 to 0.8 full at designed discharge, i.e. at the maximum estimated discharge.
- The sewage is conveyed to the point usually located in low-lying area, where the treatment plant is located.

- Treatment plant should be designed taking into consideration the quality of raw sewage expected and to meet the discharge standards.

Difference between Water Supply Pipes and Sewer Pipes

The major difference between the water distribution network and sewerage system is presented in the Table.

Table: Comparison between the water distribution network and sewage collection system

Water Supply Pipes	Sewer Pipes
• It carries pure water.	It carries contaminated water containing organic or inorganic solids which may settle in the pipe. It can cause corrosion of the pipe material.
• Velocity higher than self-cleansing is not essential, because of solids are not present in suspension.	• To avoid deposition of solids in the pipes self-cleansing velocity is necessary at all possible discharge.
• It carries water under pressure. Hence, the pipe can be laid up and down the hills and the valleys within certain limits.	• It carries sewage under gravity. Therefore it is required to be laid at a continuous falling gradient in the downward direction towards outfall point.
• These pipes are flowing full under pressure.	• Sewers are design to run partial full at maximum discharge. This extra space ensures non-pressure gravity flow. This will minimize the leakage from sewer, from the faulty joints or crack, if any.

Provision of Freeboard in Sewers

Sanitary Sewers

Sewers with diameter less than 0.4 m are designed to run half full at maximum discharge, and sewers with diameter greater than 0.4 m are designed to flow 2/3 to ¾ full at maximum discharge. The extra space provided in the sewers provides factor of safety to counteract against the following factors:

1. Safeguard against lower estimation of the quantity of wastewater to be collected at the end of design period due to private water supply by industries and public. Thus, to ensure that sewers will never flow full eliminating pressure flow inside the sewer.
2. Large scale infiltration of storm water through wrong or illegal connection, through underground cracks or open joints in the sewers.
3. Unforeseen increase in population or water consumption and the consequent increase in sewage production.

Storm Water Drains

Storm water drains are provided with nominal freeboard, above their designed full supply line because the overflow from storm water drains is not much harmful. Minimum of 0.3 m free board is generally provided in storm water drains.

Hydraulic Formulae for Determining Flow Velocities

Sewers of any shape are hydraulically designed as open channels, except in the case of inverted siphons and discharge lines of pumping stations. Following formulae can be used for design of sewers.

Manning's Formula

This is most commonly used for design of sewers. The velocity of flow through sewers can be determined using Manning's formula as below:

$$v = \frac{1}{n} r^{2/3} s^{1/2} \qquad (1)$$

Where,

v = velocity of flow in the sewer, m/sec

r = Hydraulic mean depth of flow, m = a/p

a = Cross section area of flow, m^2

p = Wetted perimeter, m

n = Rugosity coefficient, depends upon the type of the channel surface i.e., material and lies between 0.011 and 0.015. For brick sewer it could be 0.017 and 0.03 for stone facing sewers.

s = Hydraulic gradient, equal to invert slope for uniform flows.

Chezy's Formula

$$V = C\, r^{1/2} s^{1/2} \qquad (2)$$

Where, C is Chezy's constant and remaining variables are same as above equation.

Crimp and Burge's Formula

$$V = 83.5\, r^{2/3} s^{1/2} \qquad (3)$$

Hazen-Williams Formula

$$V = 0.849\, C\, R^{0.63} S^{0.54} \qquad (4)$$

The Hazen-Williams coefficient 'C' varies with life of the pipe and it has high value when the pipe is new and lower value for older pipes. For example for RCC new pipe it is 150 and the value recommended for design is 120, as the pipe interior may become rough with time. The design values of 'C; for AC pipes, Plastic pipes, CI pipes, and steel lined with cement are 120, 120, 100, and 120, respectively. Modified Hazen-William's equation is also used in practice.

Minimum Velocity: Self Cleansing Velocity

The velocity that would not permit the solids to settle down and even scour the deposited particles of a given size is called as self-cleansing velocity. This minimum velocity should at least develop once in a day so as not to allow any deposition in the sewers. Otherwise, if such deposition takes place, it will obstruct free flow causing further deposition and finally leading to the complete blocking of the sewers. This minimum velocity or self-cleansing velocity can be worked out as below:

$$Vs = \sqrt{\frac{8K}{f'}(Ss-1)g.d'} \quad (5)$$

Where,

K= constant, for clean inorganic solids = 0.04 and for organic solids = 0.06

f' = Darcy Weisbach friction factor (for sewers = 0.03)

Ss = Specific gravity of sediments

g = gravity acceleration

d' = diameter of grain, m

- Hence, for removing the impurities present in sewage i.e., sand up to 1 mm diameter with specific gravity 2.65 and organic particles up to 5 mm diameter with specific gravity of 1.2, it is necessary that a minimum velocity of about 0.45 m/sec and an average velocity of about 0.9 m/sec should be developed in sewers.
- Hence, while finalizing the sizes and gradients of the sewers, they must be checked for the minimum velocity that would be generated at minimum discharge, i.e., about 1/3 of the average discharge.
- While designing the sewers the flow velocity at full depth is generally kept at about 0.8 m/sec or so. Since, sewers are generally designed for ½ to ¾ full, the velocity at 'designed discharge' (i.e., ½ to ¾ full) will even be more than 0.8 m/sec. Thus, the minimum velocity generated in sewers will help in the following ways:
    - Adequate transportation of suspended solids,
    - Keeping the sewer size under control; and
    - Preventing the sewage from decomposition by moving it faster, thereby preventing evolution of foul gases.

Maximum Velocity or Non-scouring Velocity

The interior surface of the sewer pipe gets scored due to the continuous abrasion caused by suspended solids present in sewage. The scoring is pronounced at higher velocity than what can be tolerated by the pipe materials. This wear and tear of the sewer pipes will reduce the life span of the pipe and their carrying capacity. In order to avoid this, it is necessary to limit the maximum velocity that will be produced in sewer pipe at any time. This limiting or non- scouring velocity mainly depends upon the material of sewer. The limiting velocity for different sewer material is provided in Table.

Table: Limiting or non-scouring velocity for different sewer material

Sewer Material	Limiting velocity, m/sec
Vitrified tiles	4.5 – 5.5
Cast iron sewer	3.5 – 4.5
Cement concrete	2.5 – 3.0
Stone ware sewer	3.0 – 4.5
Brick lined sewer	1.5 – 2.5

The problem of maximum or non-scouring velocity is severe in hilly areas where ground slope is very steep and this is overcome by constructing drop manholes at suitable places along the length of the sewer.

Effect of Flow Variations on Velocities in a Sewer

The discharge flowing through sewers varies considerably from time to time. Hence, there occur variation in depth of flow and thus, variation in Hydraulic Mean Depth (H.M.D.). Due to change in H.M.D. there occur changes in flow velocity, because it is proportional to $(H.M.D.)^{2/3}$. Therefore, it is necessary to check the sewer for minimum velocity of about 0.45 m/sec at the time of minimum flow (1/3 of average flow) and the velocity of about 0.9 to 1.2 m/sec should be developed at a time of average flow. The velocity should also be checked for limiting velocity i.e. non-scouring velocity at the maximum discharge.

For flat ground sewers are designed for self-cleansing velocity at maximum discharge. This will permit flatter gradient for sewers. For mild slopping ground, the condition of developing self-cleansing velocity at average flow may be economical. Whereas, in hilly areas, sewers can be designed for self-cleansing velocity at minimum discharge, but the design must be checked for non-scouring velocity at maximum discharge.

Hydraulic Characteristics of Circular Sewer Running Full or Partially Full

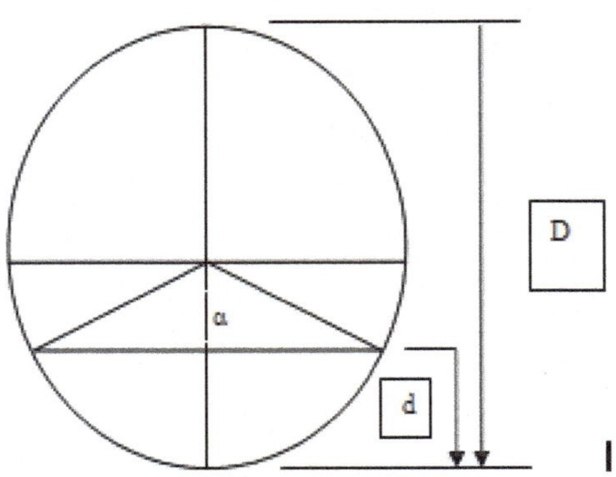

Section of a circular sewer running partially full

a) Depth at Partial flow

$$d = \left[\frac{D}{2} - \frac{D}{2}\cos\left(\frac{\alpha}{2}\right)\right] \quad (6)$$

b) Therefore proportionate depth

$$\frac{d}{D} = \frac{1}{2}\left[1 - \cos\left(\frac{\alpha}{2}\right)\right] \quad (7)$$

c) Proportionate area

$$\frac{\alpha}{A} = \left[\frac{\alpha}{360} - \frac{Sin\alpha}{2\pi}\right] \quad (8)$$

d) Proportionate perimeter: $\frac{\rho}{P} = \frac{\alpha}{360}$ (9)

e) Proportionate Hydraulic Mean Depth

$$\frac{r}{R} = \left[1 - \frac{360 Sin\alpha}{2\pi\alpha}\right] \quad (10)$$

f) Proportionate velocity = $\frac{v}{V} = \frac{N}{n}\frac{r^{2/3}}{R^{2/3}}$ (11)

In all above equations except 'α' everything is constant. Hence, for different values of 'α', all the proportionate elements can be easily calculated. These values of the hydraulic elements can be obtained from the proportionate graph prepared for different values of d/D. The value of Manning's n can be considered constant for all depths. In reality, it varies with the depth of flow and it may be considered variable with depth and accordingly the hydraulic elements values can be read from the graph for different depth ratio of flow.

From the plot it is evident that the velocities in partially filled circular sewer sections can exceed those in full section and it is maximum at d/D of 0.8. Similarly, the discharge obtained is not maximum at flow full condition, but it is maximum when the depth is about 0.95 times the full depth.

The sewers flowing with depths between 50% and 80% full need not to be placed on steeper gradients to be as self cleansing as sewers flowing full. The reason is that velocity and discharge are function of tractive force intensity which depends upon friction coefficient as well as flow velocity generated by gradient of the sewer. Using subscript 's' denoting self cleansing equivalent to that obtained in full section, the required ratios vs/V, qs/Q and ss/S can be computed as stated below:

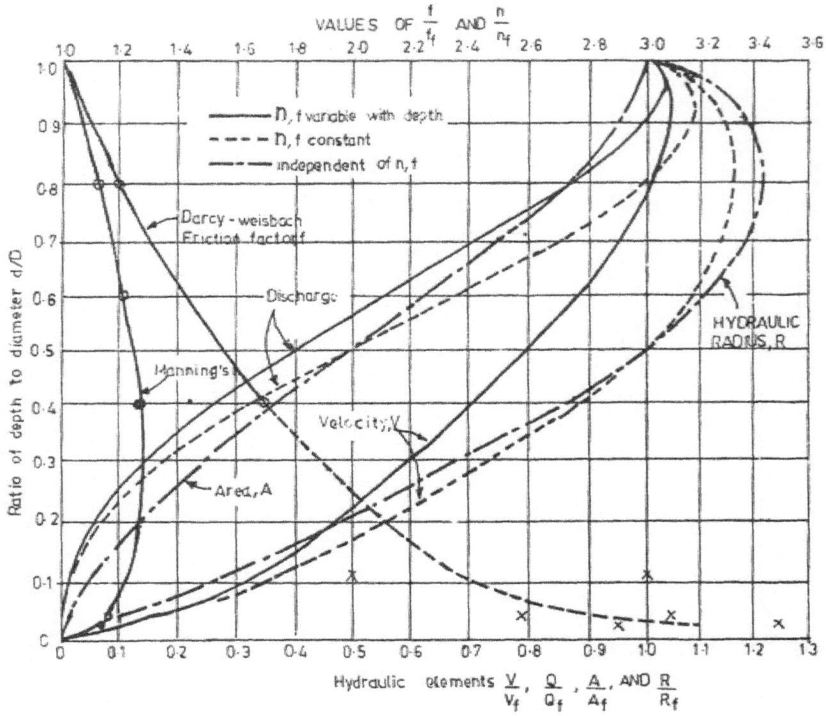

(a) Hydraulic Elements for Circular Sewer

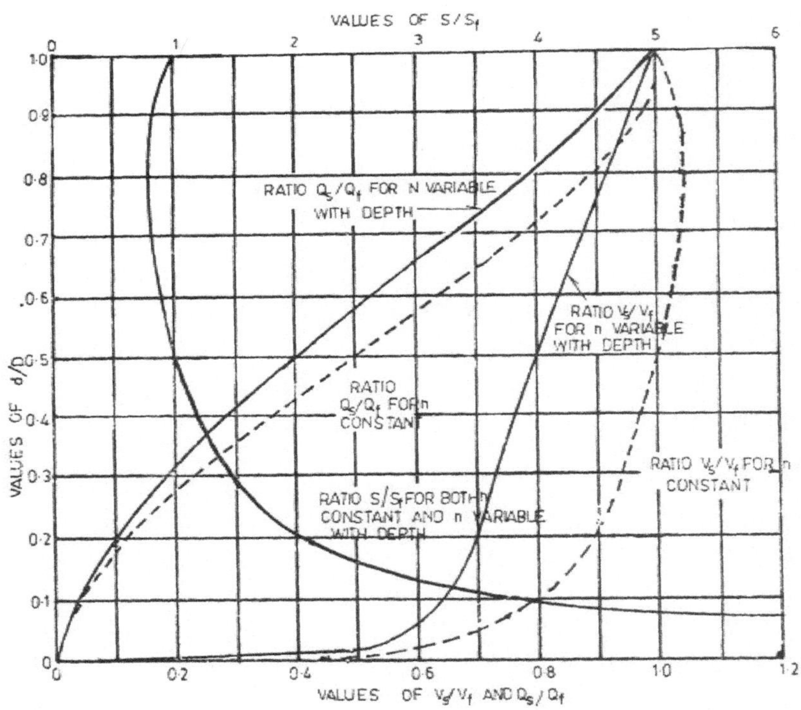

(b) Hydraulic elements of circular sewer possessing equal selfcleansing properties at all depths

Proportionate graph for circular sewer section (CPHEEO Manual, 1993)

Consider a layer of sediment of unit length, unit width and thickness 't', is deposited at the invert of the sewer. Let the slope of the sewer is θ degree with horizontal. The drag force or the intensity of tractive force (ι) exerted by the flowing water on a channel is given by:

$$\iota = \gamma_w \,.\, R.\, S \qquad (12)$$

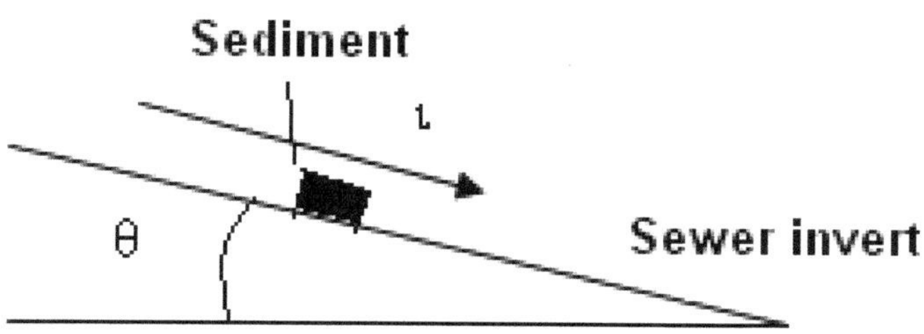

A sediment particle moving on the sewer invert

Where,

γ_w = unit weight of water

R = Hydraulic mean depth

S = slope of the invert of the sewer per unit length

With the assumption that the quantity of tractive force intensity at full flow and partial flow implies equality of cleansing, i.e., for sewers to be same self-cleansing at partial depth as full depth:

$$\iota = T$$

Therefore, $\gamma_w \,.\, r.\, ss = \gamma_w \,.\, R.\, S$ (13)

Hence, $S_s = (R/r)\, S$

Or

$$\frac{S_s}{S} = \frac{R}{r} \qquad (14)$$

Therefore,

$$\frac{V_s}{V} = \frac{N}{n}\left(\frac{r}{R}\right)^{2/3}\left(\frac{S_s}{S}\right)^{1/2} \qquad (15)$$

OR, by substituting $r/R = S/s_s$

$$\frac{\ddot{u}_s}{\ddot{u}} = -\left(-\right)^{\ddot{u}} \qquad (16)$$

And

$$\frac{q_s}{Q} = \frac{N}{n}\frac{a}{A}\left(\frac{r}{R}\right)^{1/6} \qquad (17)$$

Example: 2

A 300 mm diameter sewer is to flow at 0.3 depth on a grade ensuring a degree of self cleansing equivalent to that obtained at full depth at a velocity of 0.9 m/sec. Find the required grade and associated velocity and rate of discharge at this depth. Assume Manning's rugosity coefficient n = 0.013. The variation of n with depth may be neglected.

Solution:

Manning's formula for partial depth

$$v = \frac{1}{n} r^{2/3} s^{1/2}$$

For full depth

$$v = \frac{1}{N} R^{2/3} S^{1/2}$$

Using V = 0.90 m/sec, N = n = 0.013 and R = D/4 = 75 mm = 0.075 m

$$0.90 = \frac{1}{0.013} 0.075^{2/3} S^{1/2}$$

$$S = 0.0043$$

This is the gradient required for full depth.

and, $Q = A.V = \pi/4\ (0.3)^2 \times 0.90 = 0.064\ m^3/s$

At depth d = 0.3D, (i.e., for d/D = 0.3) we have a/A = 0.252 and r/R = 0.684 (neglecting variation of n)

Now for the sewer to be same self cleansing at 0.3 m depth as it will be at full depth, we have the gradient (S_s) required as $S_s = (R/r)S$

Therefore, $S_s = S/0.684$

$= 0.0043 / 0.0684 = 0.0063$

Now, the velocity v_s generated at this gradient is given by

$$v_s = V \frac{N}{n} \left(\frac{r}{R} \right)^{1/6}$$

$$= 1 \times (0.684)^{1/6} \times 0.9$$

$$= 0.846 \text{ m/s}$$

The discharge q_s is given by

$$q_s = Q \frac{N}{n} \frac{a}{A} \left(\frac{r}{R} \right)^{1/6}$$

q_s = 1 x (0.258) x (0.939) x (0.064)

= 0.015 m^3/s

Example: 3

A combined sewer was designed to serve an area of 60 sq. km with an average population density of 185 persons/hectare. The average rate of sewage flow is 350 L/Capita/day. The maximum flow is 50% in excess of the average sewage flow. The rainfall equivalent of 12 mm in 24 h can be considered for design, all of which is contributing to surface runoff. What will be the discharge in the sewer? Find the diameter of the sewer if running full at maximum discharge.

Solution:

Total population of the area = population density x area

= 185 x 60 x 10^2

= 1110 x 10^3 persons

Average sewage flow = 350 x 11.1 x 10^5 Liters/day

= 388.5 x 10^6 L/day

= 4.5 m^3/sec

Storm water flow = 60 x 10^6 x (12/1000) x [1/(24 x 60 x 60)]

= 8.33 m^3/sec

Maximum sewage flow = 1.5 x average sewage flow

= 1.5 x 4.5 = 6.75 m^3/sec

Total flow of the combined sewer = sewage flow + storm flow

= 6.75 + 8.33 = 15.08 m^3/sec

Hence, the capacity of the sewer = 15.08 m^3/sec

Hence, diameter of the sewer required at the velocity of 0.9 m/s can be calculated as

$$\pi/4\,(D)^2 \times 0.90 = 15.08\ m^3/s$$

Hence, D = 4.62 m

Example: 4

Find the minimum velocity and gradient required to transport coarse sand through a sewer of 40 cm diameter with sand particles of 1.0 mm diameter and specific gravity 2.65, and organic matter of 5 mm average size with specific gravity 1.2. The friction factor for the sewer material may be assumed 0.03 and roughness coefficient of 0.012. Consider k = 0.04 for inorganic solids and 0.06 for organic solids.

Solution

Minimum velocity i.e. self cleansing velocity

$$Vs = \sqrt{\frac{8k}{f'}(Ss-1)gd'}$$

$$Vs = \sqrt{\frac{8x0.04}{0.03}(2.65-1)x9.81x0.001}$$

= 0.4155 m/sec say 0.42 m/sec

Similarly, for organic solids this velocity will be 0.396 m/sec

Therefore, the minimum velocity in sewer = 0.42 m/sec

Now, Diameter of the sewer D = 0.4 m

Hydraulic Mean Depth = D/4 = 0.4/4 = 0.1 m

Using Manning's formula:

$$V = 1/n\ R^{2/3}\ S^{1/2}$$

$$0.42 = (1/0.012) \times (0.1)^{2/3} \times S^{1/2}$$

$$S = 1/1824.5$$

Therefore, gradient of the sewer required is 1 in 1824.5.

Design of Storm Water Drains for Separate System

Important points for design

Storm water is collected from streets into the link drains, which in turn discharge into main drains of open type. The main drain finally discharges the water into open water body. As far as possible gravity discharge is preferred, but when it is not possible, pumping can be employed. While designing, the alignment of link drains, major drains and sources of disposal are properly planned on contour maps. The maximum discharge expected in the drains is worked out. The longitudinal sections of the drains are prepared keeping in view the full supply level (FSL) so that at no place it should go above the natural surface level along the length. After deciding the FSL line, the bed line is fixed (i.e. depth of drain) based on following consideration.

a. The bed level should not go below the bed level of source into which storm water is discharged.

b. The depth in open drain should preferably be kept less than man height.

c. The depth is sometimes also decided based on available width.

d. The drain section should be economical and velocities generated should be non-silting and non-scouring in nature. The drain section is finally designed using Manning's formula. Adequate free board is provided over the design water depth at maximum discharge.

Laying of Sewer Pipes

- Sewers are generally laid starting from their outfall ends towards their starting points. With this advantage of utilization of the tail sewers even during the initial periods of its construction is possible.
- It is common practice, to first locate the points where manholes are required to be constructed as per drawing, i.e., L-section of sewer, and then laying the sewer pipe straight between the two manholes.
- The central line of the sewer is marked on the ground and an offset line is also marked parallel to the central line at suitable distance, about half the trench width plus 0.6 m. This line can be drawn by fixing the pegs at 15 m intervals and can be used for finding out center line of the sewer simply by offsetting.
- The trench of suitable width is excavated between the two manholes and the sewer is laid between them. Further excavation is then carried out for laying the pipes betweenthe next consecutive manholes. Thus, the process is continued till the entire sewers are laid out.
- The width of the trench at the bottom is generally kept 15 cm more than the diameter of the sewer pipe, with minimum 60 cm width to facilitate joining of pipes.
- If the sewer pipes are not to be embedded in concrete, such as for firm grounds, then the bottom half portion of the trench is excavated to confirm the shape of the pipe itself. In ordinary or softer grounds, sewers are laid embedded in concrete.
- The trench is excavated up to a level of the bottom embedding concrete or up to the invert level of the sewer pipe plus pipe thickness if no embedding concrete is provided. The designed invert levels and desired slope as per the longitudinal section of the sewer should be precisely transferred to the trench bottom.
- After bedding concrete is laid in required alignment and levels. The sewer pipes are then lowered down into the trench either manually or with the help of machines for bigger pipe diameters.
- The sewer pipe lengths are usually laid from the lowest point with their sockets facing up the gradient, on desired bedding. Thus, the spigot end of new pipe can be easily inserted on the socket end of the already laid pipe.

Hydraulic Testing of Sewers

Test for Leakage or Water Test

The sewers are tested after giving sufficient time for the joints to set for no leakage. For this sewer pipe sections are tested between the manholes to manhole under a test pressure of about 1.5 m water head. To carry this, the downstream end of the sewer is plugged and water is filled in the manhole at upper end. The depth of water in manhole is maintained at about 1.5 m. The sewer line is inspected and the joints which leak are repaired.

Test for Straightness of Alignment

This can be tested by placing a mirror at one end of the sewer line and a lamp at the other end. If the pipe line is straight, full circle of light will be observed.

Backfilling the trench: After the sewer line has been laid and tested, the trenches are back filled. The earth should be laid equally on either side with layer of 15 cm thickness. Each layer should be properly watered and rammed.

Manhole

A manhole (alternatively utility hole, cable chamber, maintenance hole, inspection chamber, access chamber, sewer hole, or confined space) is the top opening to an underground utility vault used to house an access point for making connections, inspection, valve adjustments or performing maintenance on underground and buried public utility and other services including water, sewers, telephone, electricity, storm drains, district heating and gas.

PMG manholes in a city street, Perth, Western Australia.

Manhole in Belo Horizonte, Brazil.

Manhole cover of precast concrete in Germany

Manhole being used to access sewer

Usage

Manhole closings are protected by a manhole cover, a flat plug designed to prevent accidental or unauthorized access to the manhole. Those plugs are traditionally made of metal, but may be constructed from precast concrete, glass reinforced plastic or other composite material (especially in Europe, or where cover theft is of concern).

Manholes are usually outfitted with metal, polypropylene, or fiberglass steps installed in the inner side of the wall to allow easy descent into the utility space. Because of legislation restricting acceptable manual handling weights, Europe has seen a move toward lighter weight composite manhole cover materials, which also have the benefits of greater slip resistance and electrical insulating properties.

The access openings are usually circular in shape to prevent accidental fall of the cover into the hole.

Manholes are generally found in urban areas, in streets and occasionally under sidewalks. In rural and undeveloped areas, services such as telephone and electricity are usually carried on utility poles or even pylons rather than underground.

Composite Manholes

Composite (fiberglass) manholes are commonly used in applications where infiltration, exfiltration, or corrosion by hydrogen sulfide (from sewer gas) are a concern, or where structures need to be factory integrated into a manhole before placement.

Structures commonly integrated into composite manholes include:

- Flow inverts
- Flumes
- Drop structures from higher elevation flows to lower elevation discharge pipes

Occasionally, composite manholes will integrate:

- Weirs
- Storm water screening structures
- Sewage grinders
- Energy absorbing structures to dissipate undesirable flow stream turbulence or velocity

Hazards Caused by Stray Voltage in Manholes

In urban areas, stray voltage issues have become a significant concern for utilities. In 2004, Jodie S. Lane was electrocuted after stepping on a metal manhole cover, while walking her dog in New York City.

Sewer Appurtenances

Manholes

The manhole is masonry or R.C.C. chamber constructed at suitable intervals along the sewer lines, for providing access into them. Thus, the manhole helps in inspection, cleaning and maintenance of sewer. These are provided at every bend, junction, change of gradient or change of diameter of the sewer. The sewer line between the two manholes is laid straight with even gradient. For straight sewer line manholes are provided at regular interval depending upon the diameter of the sewer. The spacing of manhole is recommended in IS 1742-1960. For sewer up to 0.3 m diameter or sewers which cannot be entered for cleaning or inspection the maximum spacing between the manholes recommended is 30 m, and 300 m spacing for pipe greater than 2.0 m diameter. A spacing allowance of 100 m per 1 m diameter of sewer is a general rule in case of very large sewers (CPHEEO, 1993). The internal dimensions required for the manholes are provided (CPHEEO, 1993). The minimum width of the manhole should not be less than internal diameter of the sewer pipe plus 150 mm benching on both the sides.

Table: Spacing of Manholes

Pipe Diameter	Spacing
Small sewers	45 m
0.9 to 1.5 m	90 to 150 m
1.5 to 2.0 m	150 to 200 m
Greater than 2.0 m	300 m

Table: The minimum internal dimensions for manhole chambers

Depth of sewer	Internal dimensions
0.9 m or less depth	0.90 m x 0.80 m
For depth between 0.9 m and 2.5 m	1.20 m x 0.90 m, 1.2 m dia. for circular
For depth above 2.5 m and up to 9.0 m	For circular chamber 1.5 m dia.
For depth above 9.0 m and up to 14.0 m	For circular chamber 1.8 m dia.

Classification of Manholes

Depending upon the depth the manholes can be classified as:

(a) Shallow Manholes, (b) Normal Manholes, and (c) Deep Manholes

Shallow Manholes: These are 0.7 to 0.9 m depth, constructed at the start of the branch sewer or at a place not subjected to heavy traffic conditions. These are provided with light cover at top and called inspection chamber.

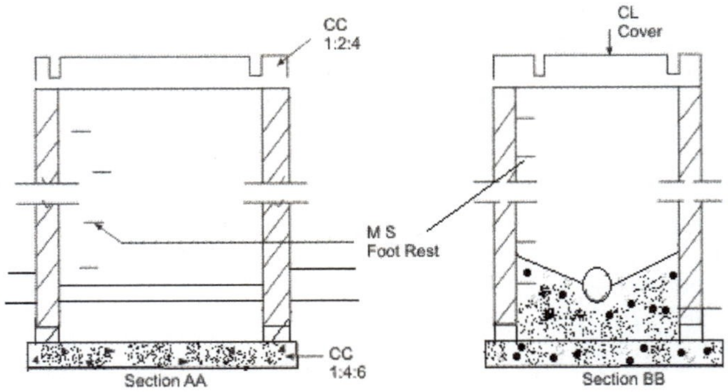

Shallow manhole

Normal Manholes: These manholes are 1.5 m deep with dimensions 1.0 m x 1.0 m square or rectangular with 1.2 m x 0.9 m. These are provided with heavy cover at its top to support the anticipated traffic load.

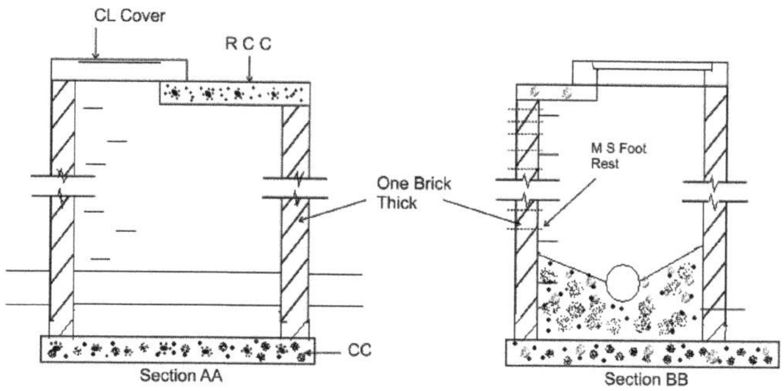

Rectangular manhole for depth 0.9 m to 2.5 m

Deep Manholes: The depth of these manholes is more than 1.5 m. The section of such manhole is not uniform throughout. The size in upper portion is reduced by providing an offset. Steps are provided in such manholes for descending into the manhole. These are provided with heavy cover at its top to support the traffic load.

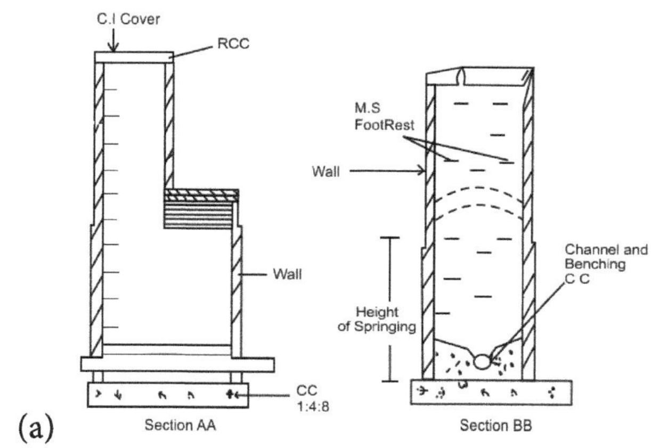

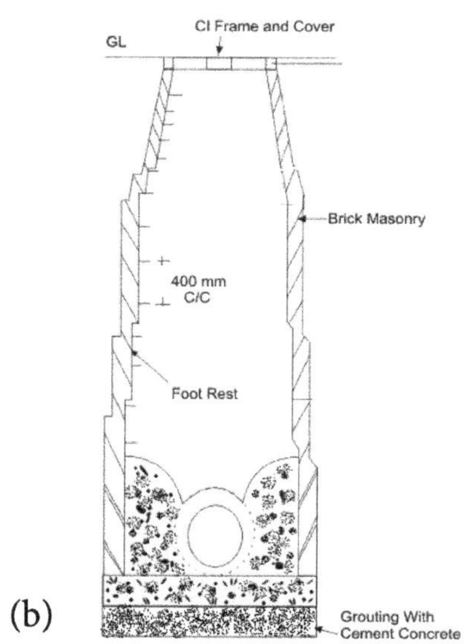

(a) Rectangular and (b) Circular deep manhole

Other Types of Manholes

Straight – Through Manholes

This is the simplest type of manhole, which is built on a straight run of sewer with no side junctions. Where there is change in the size of sewer, the soffit or crown level of the two sewers should be the same, except where special conditions require otherwise.

Junction Manholes

This type of manholes are constructed at every junction of two or more sewers, and on the curved portion of the sewers, with curved portion situated within the manhole. This type of manholes can be constructed with the shape other than rectangular to suit the curve requirement and achieve economy. The soffit of the smaller sewer at junction should not be lower than that of the larger sewer. The gradient of the smaller sewer may be made steeper from the previous manhole to reduce the difference of invert at the point of junction to a convenient amount.

Side Entrance Manholes

In large sewers where it is difficult to obtain direct vertical access to the sewer from the top ground level due to obstructions such as, other pipe lines like water, gas, etc., the access shaft should be constructed in the nearest convenient position off the line of sewer, and connected to the manhole chamber by a lateral passage. The floor of the side entrance passage which should fall at about 1 in 30 towards the sewer should enter the chamber not lower than the soffit level of the sewer. In large sewers necessary steps or a ladder (with safety chain or removable handrail) should be provided to reach the benching from the side entrance above the soffit.

Drop Manholes

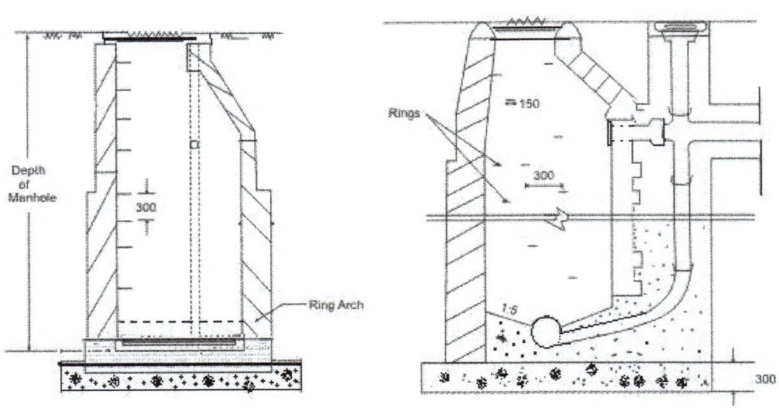

Drop manhole

When a sewer connects with another sewer, where the difference in level between invert level of branch sewer and water line in the main sewer at maximum discharge is greater than 0.6 m, a manhole may be built either with vertical or nearly vertical drop pipe from higher sewer to the lower one. The drop manhole is also required in the same sewer line in sloping ground, when drop more than 0.6 m is required to control the gradient and to satisfy the maximum velocity i.e., non-scouring velocity.

The drop pipe may be outside the shaft and encased in concrete or supported on brackets inside the shaft. If the drop pipe is outside the shaft, a continuation of the sewer should be built through the shaft wall to form a rodding and inspection eye, provided with half blank flange. When the drop pipe is inside the shaft, it should be of cast iron and provided with adequate arrangements for rodding and with water cushion of 150 mm depth at the end. The diameter of the drop pipe should be at least equal to incoming pipe.

Scraper (Service) Type Manhole

All sewers above 450 mm in diameter should have one manhole at intervals of 110 to 120 m of scraper type. This manhole should have clear opening of 1.2 m × 0.9 m at the top to facilitate lowering of buckets.

Flushing Manholes

In flat ground for branch sewers, when it is not possible to obtain self cleansing velocity at all flows, due to very little flow, it is necessary to incorporate flushing device. This is achieved by making grooves at intervals of 45 to 50 m in the main drains in which wooden planks are inserted and water is allowed to head up. When the planks are removed, the water will rush with high velocity facilitating cleaning of the sewers. Alternatively, flushing can be carried out by using water from overhead water tank through pipes and flushing hydrants or through fire hydrants or tankers and hose.

Flushing manholes are provided at the head of the sewers. Sufficient velocity shall be imparted in the sewer to wash away the deposited solids. In case of heavy chocking in sewers, care should be exercised to ensure that there is no possibility of back flow of sewage into the water supply mains.

Inverted Siphons

An inverted siphon or depressed sewer is a sewer that runs full under gravity flow at a pressure above atmosphere in the sewer. Inverted siphons are used to pass under obstacles such as buried pipes, subways, etc. This terminology 'siphon' is misnomer as there is no siphon action in the depressed sewer. As the inverted siphon requires considerable attention for maintenance, it should be used only where other means of passing an obstacle in line of the sewer are impracticable.

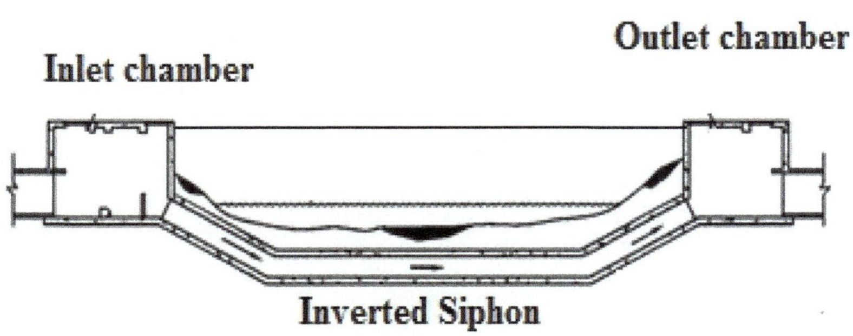

Inverted siphon

Stormwater Inlets

Storm water inlets are provided to admit the surface runoff to the sewers. These are classified in three major groups viz. curb inlets, gutter inlets, and combined inlets. They are provided either depressed or flush with respect to the elevation of the pavement surface. The structure of the inlet is constructed with brickwork with cast iron grating at the opening confirming to IS 5961. Where the traffic load is not expected, fabricated steel grating can be used. The clear opening shall not be more than 25 mm. The connecting pipe from the street inlet to the sewer should be minimum of 200 mm diameter and laid with sufficient slope. A maximum spacing of 30 m is recommended between the inlets, which depends upon the road surface, size and type of inlet and rainfall.

Curb Inlet: These are vertical opening in the road curbs through which stormwater flow enters the stormwater drains. These are preferred where heavy traffic is anticipated.

Gutter Inlets: These are horizontal openings in the gutter which is covered by one or more grating through which stormwater is admitted.

Combined Inlets: In this, the curb and gutter inlet both are provided to act as a single unit. The gutter inlet is normally placed right in front of the curb inlets.

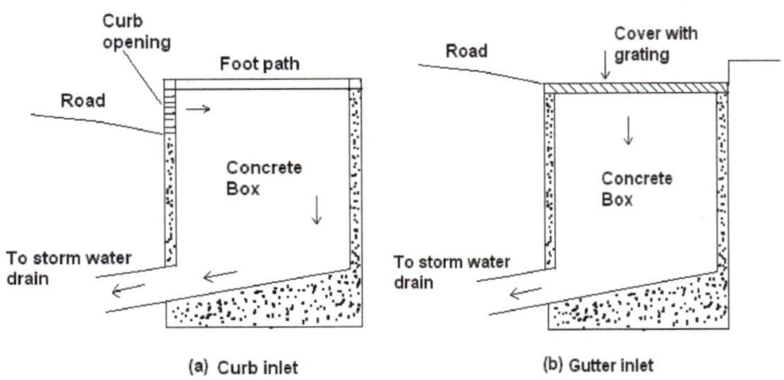

(a) Curb inlet and (b) Gutter inlet

Trap (Plumbing)

Water seal in drain pipe under a sink. Water enters at right, fills the trap, and continues left. Siphoning takes place above the line "B", while inverted siphoning occurs below the line "A".

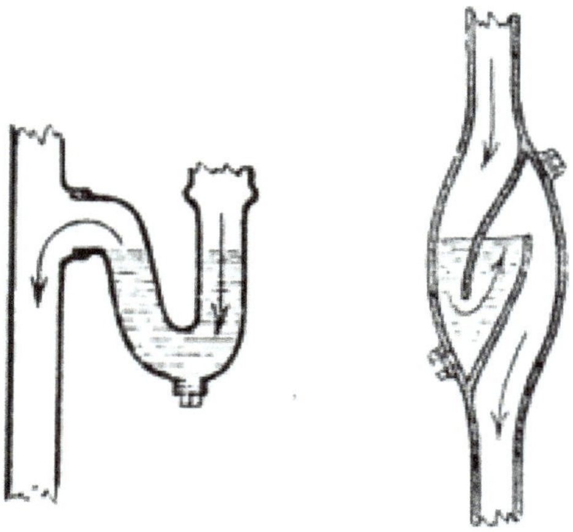

Examples of traps

In plumbing, a trap is a U, S, or J-shaped pipe located below or within a plumbing fixture. An S-shaped trap is also known as an S-bend. It was invented by Alexander Cummings in 1775 but became known as the U-bend following the introduction of the U-shaped trap by Thomas Crapper in 1880. The new U-bend could not jam, so, unlike the S-bend, it did not need an overflow. The bend is used to prevent sewer gases from entering buildings. In refinery applications, it also prevents hydrocarbons and other dangerous gases from escaping outside through drains.

The most common of these traps in houses is referred to as a P-trap. It is the addition of a 90 degree fitting on the outlet side of a U-bend, thereby creating a P-like shape. It is also referred to as a sink trap because it is installed under most house sinks.

Because of its shape, the trap retains a small amount of water after the fixture's use. This water in the trap creates a seal that prevents sewer gas from passing from the drain pipes back into the occupied space of the building. Essentially all plumbing fixtures including sinks, bathtubs, and toilets must be equipped with either an internal or external trap.

Because it is a localized low-point in the plumbing, sink traps also tend to capture heavy objects (such as jewelry) that are inadvertently dropped into the sink. Traps also tend to collect hair, sand, and other debris and limit the ultimate size of objects that will pass on into the rest of the plumbing, thereby catching oversized objects. For all of these reasons, most traps can either be disassembled for cleaning or they provide some sort of cleanout feature.

When a large volume of water may be discharged through the trap, a standpipe may be required to prevent impact to other nearby traps.

Venting and Auxiliary Devices

Maintaining the water seal is critical to trap operation; traps can and do dry out, and poor venting can suction or blow water out of the traps. This is usually avoided by venting the drain pipes downstream of the trap; by being vented to the atmosphere outside the building, the drain lines never operate at a pressure much higher or lower than atmospheric pressure. Plumbing codes usually provide strict limitations on how far a trap may be located from the nearest *vent stack*.

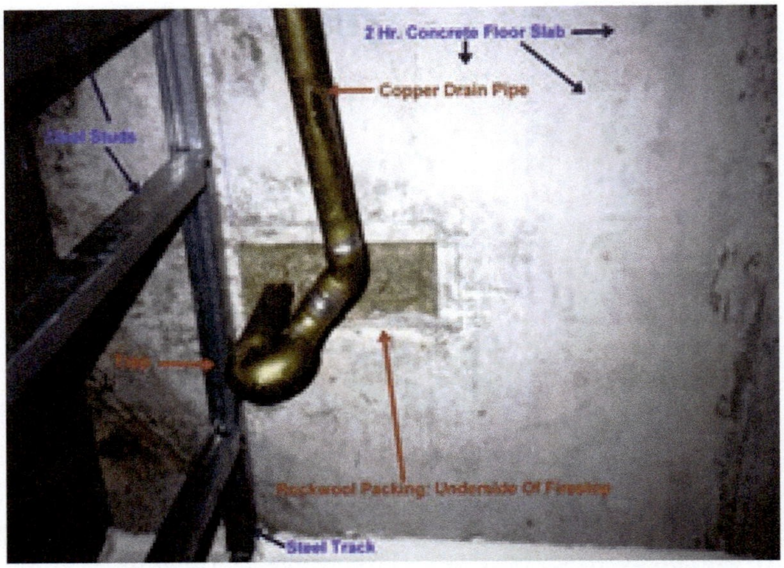

Trap with copper drain pipe at underside of firestop packing in 2 hour fire-resistance rated concrete floor slab

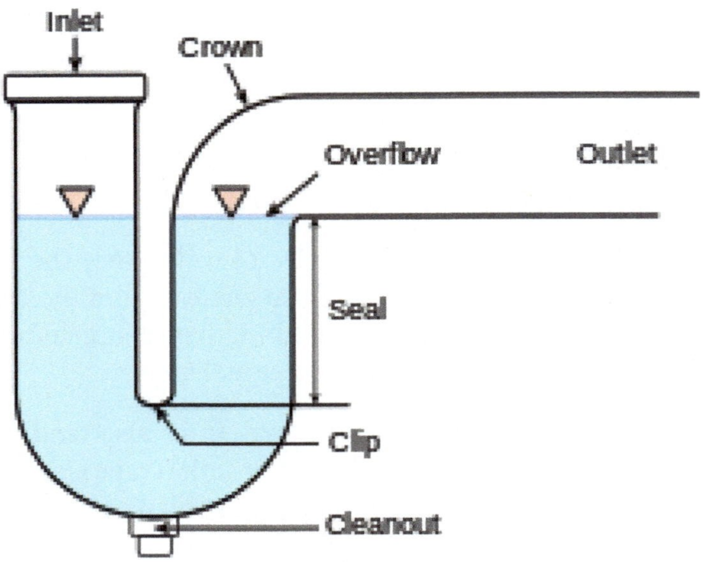

Typical P-trap

When a vent cannot be provided, codes may allow the use of an air admittance valve instead. These devices avoid negative pressure in the drain pipe by venting room air into the drain pipe (behind the trap). A "Chicago Loop" is another alternative.

When a trap is installed on a fixture that is not routinely used, the eventual evaporation of the water in the trap must be considered. In these cases, a trap primer may be installed; these are devices that automatically recharge traps with water to maintain their water seals.

Accepted Traps

In some regions of the United States, "S" traps are no longer accepted by the plumbing codes or are even illegal, as these traps tend to easily siphon dry even when well-vented. It is often possible to tell whether a household uses an S- or U-bend by checking for the presence of an overflow pipe outlet. Certain drum-styled traps are also discouraged or banned.

Grease Trap

Grease trap for greywater in Lima, Peru

A grease trap (also known as grease interceptor, grease recovery device and grease converter) is a plumbing device designed to intercept most greases and solids before they enter a wastewater disposal system. Common wastewater contains small amounts of oils which enter into septic tanks and treatment facilities to form a floating scum layer. This scum layer is very slowly digested and broken down by microorganisms in the anaerobic digestion process. Large amounts of oil from food preparation in restaurants can overwhelm a septic tank or treatment facility, causing release of untreated sewage into the environment. High-viscosity fats and cooking grease such as lard solidify when cooled, and can combine with other disposed solids to block drain pipes.

Grease traps have been used since Victorian days. Nathaniel Whiting obtained the first patent for a grease trap in the late 1800s. These reduce the amount of fats, oils and greases (FOGs) that enter sewers. They are boxes within the drain run that flows between the sinks in a kitchen the sewer system. They only have kitchen waste water flowing through them, and do not serve any other drainage system, such as toilets. They can be made from many different materials, such as stainless steel, plastics, concrete & cast iron. They range from 35 liter capacity to 45,000 liters and greater. They can be located above ground, below ground, inside the kitchen or outside the building.

Types

There are three primary types of devices. The most common are the types specified by ASME (American Society Of Mechanical Engineers), utilizing baffles, or a proprietary inlet diffuser.

Grease trap sizing is based on the size of the 2- or 3-compartment sink, dishwasher, pot sinks, and mop sinks. The cumulative flow rates of these devices, as well as overall grease retention capacity (in pounds or kilograms) are considered. Currently, ASME Standard (ASME A112.14.3) is being adopted by both of the National Model Plumbing Codes that cover most of the US. This standard requires that grease interceptors remove a minimum of 90% of incoming FOGS (fats, oils and greases). It also requires that grease interceptors are third-party tested and certified to 90 days compliance with the standard pumping. This third-party testing must be conducted by a recognized and approved third-party testing laboratory.

Manhole covers of a grease trap outside of a restaurant.

The most common passive grease traps are smaller, point-of-use units used under three-compartment sinks or adjacent to dishwashers in the kitchen. There has been little innovation in passive grease trap technology until recently, with the introduction of grease trap liners, which provide increased health, hygiene and safety benefits for end users. They restrict flow and remove 85–90% of the incoming FOG. Food solids along with fats, oils, and grease are trapped and stored in these devices. These units are available in many sizes, are often economical, and some can accommodate substantial flow volume. For some applications, this allows a food business to be created where otherwise an in-ground solution would be cost-prohibitive.

The second most common type of interceptor is the large in-ground tank, which is usually 500–2,000 US gallons (1,900–7,600 l; 420–1,670 imp gal). These units are constructed of concrete, fiberglass, or steel. By nature of their larger size, they have larger grease and solid storage capacities for high-flow applications such as a restaurant or hospital store. They are commonly called gravity interceptors. Interceptors require a retention time of 30 minutes to allow the fats, oils, grease and food solids to settle in the tank. As more waste water enters the tank, the grease-free water is pushed out of the tank. The rotting brown grease inside a grease trap or grease interceptor must be pumped out on a scheduled basis. The brown grease is not recycled and goes to landfill. On average 300 to 400 pounds (140 to 180 kg) of brown grease goes to landfill annually from each restaurant.

A third system type, GRDs (grease recovery devices), removes the grease automatically when trapped. The recovered grease or "yellow grease" is recycled with the waste vegetable oil from the kitchen's deep-fryers. Restaurants must not pay for grease trap pumping as do restaurants with conventional grease traps or grease interceptors.

Passive grease traps and passive grease interceptors must be emptied and cleaned when 25% full. As the passive devices fill with fats, oils, and grease, they become less productive for grease recovery. A full grease trap does not stop any FOG from entering the sanitary sewer system. The emptied contents or "brown grease" is considered hazardous waste in many jurisdictions. "Brown grease" is generally not recycled and goes to landfill waste.

Uses

In-ground grease trap outside of a shop

Restaurant and foodservice kitchens produce much waste grease which is present in the drain lines from various sinks, dishwashers and cooking equipment such as combi ovens and commercial woks. If not removed, the grease can clump and cause blockage and back-up in the sewer.

In the US, sewers back up annually "an estimated 400,000 times, and municipal sewer overflows on 40,000 occasions". The EPA has determined that sewer pipe blockages are the leading cause of sewer overflows, and grease is the primary cause of sewer blockages. Even if accumulated FOG does not escalate into blockages and sanitary sewer overflows, it can disrupt wastewater utility operations and increase operations and maintenance requirements"

For these reasons, depending on the country, nearly all municipalities require commercial kitchen operations to use some type of interceptor device to collect grease before it enters sewers. Where FOG is a concern in the local wastewater system, communities have established inspection programs to ensure that these grease traps and/or interceptors are being routinely maintained.

It is estimated 50% of all sewer overflows are caused by grease blockages, with over 10 billion US gallons (3.8×10^{10} l; 8.3×10^{9} imp gal) of raw sewage spills annually.

Method of Operation

When the outflow from the kitchen sink enters the grease trap, the solid food particles sink to the bottom, while lighter grease and oil floats to the top. The relatively grease-free water is then fed into the normal septic system. The food solids at the bottom and floating oil and grease must be periodically removed in a manner similar to septic tank pumping. A traditional grease trap is not a food disposal unit. Unfinished food must be scraped into the garbage or food recycling bin. Milkshakes, gravy, sauces and food solids must be scraped off dishes before entering the sink or dishwasher.

To maintain some degree of efficiency, there has been a trend to specify larger traps. Unfortunately, providing a large tank for the effluent to stand also means that food waste has time to settle to the bottom of the tank, reducing available volume and adding to clean-out problems. Also, rotting food contained within an interceptor breaks down, producing toxic waste (such as sulfur gases) - hydrogen sulfide combines with the water present to create sulfuric acid. This attacks mild steel and concrete materials, resulting in "rot out", On the other hand, polyethylene has acid-resisting properties. A bigger interceptor is not a better interceptor. In most cases, multiple interceptors in series will separate grease much better.

Because it has been in the trap for some time, grease thus collected will be contaminated and is unsuitable for further use. This type of grease is called brown grease.

Disposal of Trapped Grease

Waste from passive grease traps and gravity interceptors is called brown grease. Brown grease is rotted food solids in combination with fats, oils, and grease (FOG). Brown grease is pumped from the traps and interceptors by grease pumping trucks. Unlike the collected yellow grease, the majority of brown grease goes to landfill sites. New facilities (2012) and new technology are beginning to allow brown grease to be recycled.

Catch Basins

Catch basins are provided to stop the entry of heavy debris present in the storm water into the sewers. However, their use is discouraged because of the nuisance due to mosquito breeding apart from posing substantial maintenance problems. At the bottom of the basin space is provided for the accumulation of impurities. Perforated cover is provided at the top of the basin to admit rain water into the basin. A hood is provided to prevent escape of sewer gas.

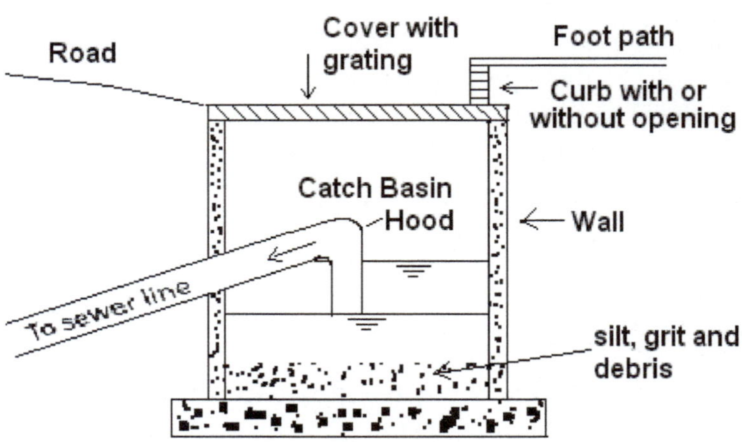

Catch basin

Clean-Outs

It is a pipe which is connected to the underground sewer. The other end of the clean-out pipe is brought up to ground level and a cover is placed at ground level. A clean-out is generally provided at the upper end of lateral sewers in place of manholes. During blockage of pipe, the cover is taken out and water is forced through the clean-out pipe to lateral sewers to remove obstacles in the sewer line. For large obstacles, flexible rod may be inserted through the clean-out pipe and moved forward and backward to remove such obstacle.

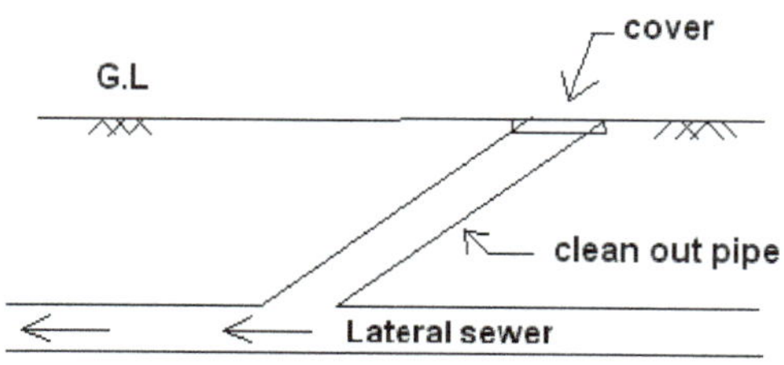

Clean-out

Regulator or Overflow Device

These are used for preventing overloading of sewers, pumping stations, treatment plants or disposal arrangement, by diverting the excess flow to relief sewer. The overflow device may be side flow or leaping weirs according to the position of the weir, siphon spillways or float actuated gates and valves.

Side Flow Weir

It is constructed along one or both sides of the combined sewer and delivers the excess flow during storm period to relief sewers or natural drainage courses. The crest of the weir is set at an elevation corresponding to the desired depth of flow in the sewer. The weir length must be sufficient long for effective regulation of the flow.

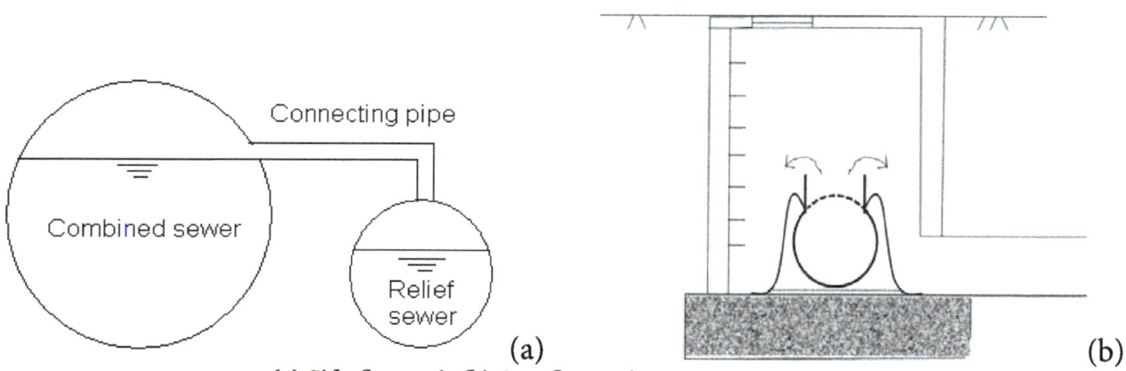

(a) Side flow weir (b) Overflow weir arrangement

Leaping Weir

The term leaping weir is used to indicate the gap or opening in the invert of a combined sewer. The leaping weir is formed by a gap in the invert of a sewer through which the dry weather flow falls and over which a portion of the entire storm leaps. This has an advantage of operating as regulator without involving moving parts. However, the disadvantage of this weir is that, the grit material gets concentrated in the lower flow channel. From practical consideration, it is desirable to have moving crests to make the opening adjustable. When discharge is small, the sewage falls directly into the intercepting sewer through the opening. But when the discharge exceeds a certain limit, the excess sewage leaps or jumps across the weir and it is carried to natural stream or river. This arrangement is shown in the Figure.

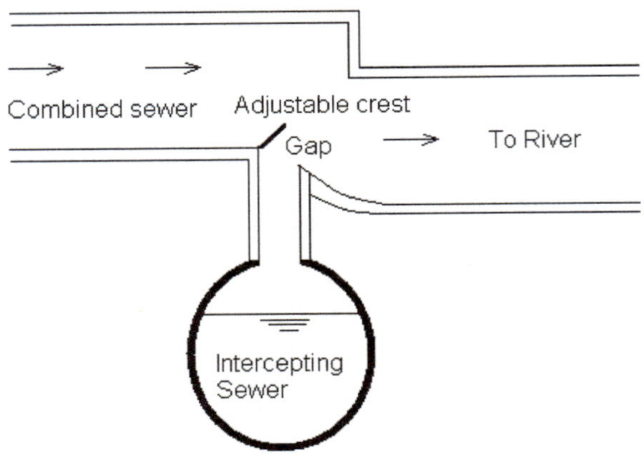

Leaping weir with adjustable crest

Float Actuated Gates and Valves

The excess flow in the sewer can also be regulated by means of automatic mechanical regulators. These are actuated by the float according to the water level in the sump interconnected to the sewers. Since, moving part is involved in this, regular maintenance of this regulator is essential.

Siphon Spillway

This arrangement of diverting excess sewage from the combined sewer is most effective because it works on the principle of siphon action and it operates automatically. The overflow channel is connected to the combined sewer through the siphon. An air pipe is provided at the crest level of siphon to activate the siphon when water will reach in the combined sewer at stipulated level.

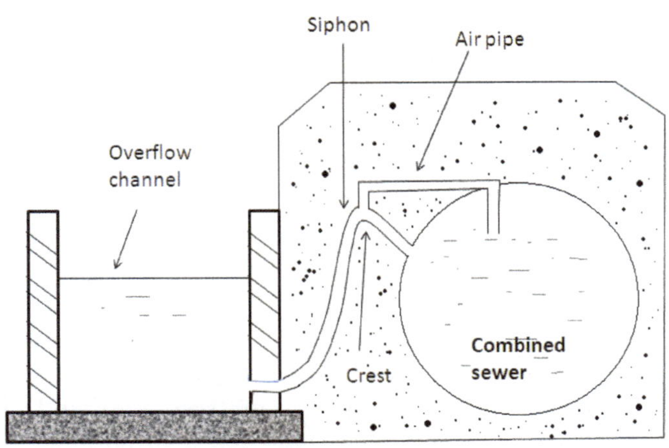

Siphonspillway

Flap Gates and Flood Gates

Flap gates or backwater gates are installed at or near sewer outlets to prevent back flow of water during high tide, or at high stages in the receiving stream. These gates can be rectangular or circular in shape and made up of wooden planks or metal alloy sheets. Such gates should be designed such that the flap should get open at a very small head difference. Adequate storage in outfall sewer is also necessary to prevent back flow into the system due to the closure of these gates at the time of high tides, if pumping is to be avoided.

Sewer Ventilators

Ventilation to the sewer is necessary to make provision for the escape of air to take care of the exigencies of full flow and to keep the sewage as fresh as possible. In case of stormwater, this can be done by providing ventilating manhole covers. In modern sewerage system, provision of ventilators is not necessary due to elimination of intercepting traps in the house connections allowing ventilation.

Lamp Hole

It is an opening or hole constructed in a sewer for purpose of lowering a lamp inside it. It consists of stoneware or concrete pipe, which is connected to sewer line through a T-junction as shown in the Figure. The pipe is covered with concrete to make it stable. Manhole cover of sufficient strength is provided at ground level to take the load of traffic. An electric lamp is inserted in the lamp hole and the light of lamp is observed from manholes. If the sewer length is unobstructed, the light of lamp will be seen. It is constructed when construction of manhole is difficult. In present practice as far as possible the use of lamp hole is avoided. This lamp hole can also be used for flushing the sewers. If the top cover is perforated it will also help in ventilating the sewer, such lamp hole is known as fresh air inlet.

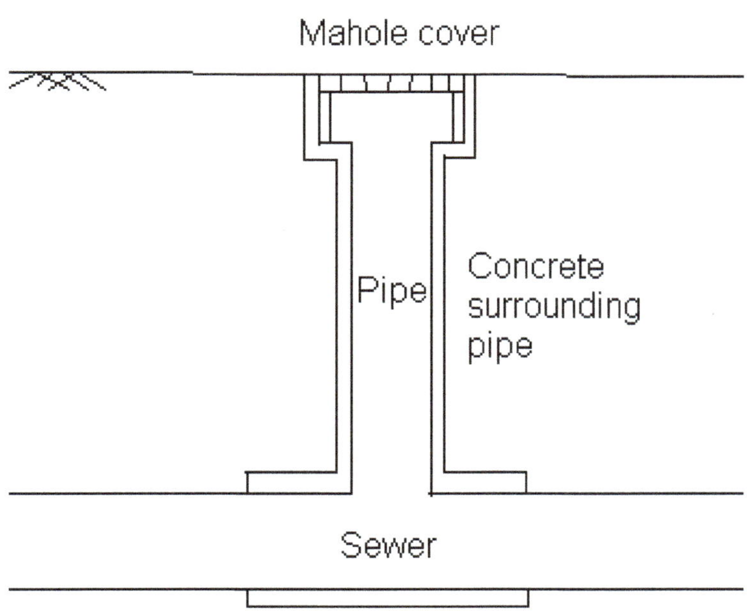

Lamp hole

Pumping Station

There are certain locations where it is possible to convey sewage by gravity to a central treatment facility or stormwater is conveyed up to disposal point entirely by gravity. Whereas, in case of large area being served with flat ground, localities at lower elevation or widely undulating topography it may be essential to employ pumping station for conveyance of sewage to central treatment plant. Sewage and stormwater is required to be lifted up from a lower level to a higher level at various places in a sewerages system. Pumping of sewage is also generally required at the sewage treatment plant.

Pumping of sewage is different than water pumping due to polluted nature of the wastewater containing suspended solids and floating solids, which may clog the pumps. The dissolved organic and inorganic matter present in the sewage may chemically react with the pump and pipe material and can cause corrosion. The disease causing bacteria present in the sewage may pose health hazard to the workers. Sedimentation of organic matter in the sump well may lead to decomposition and spreading of foul odour in the pumping station, requiring proper design to avoid deposition. Also, variation of sewage flow with time makes it a challenging task.

Pumping stations are often required for pumping of (1) untreated domestic wastewater, (2) stormwater runoff, (3) combined domestic wastewater and stormwater runoff, (4) sludge at a wastewater treatment plant, (5) treated domestic wastewater, and (6) recycling treated water or mixed liquor at treatment plants. Each pumping application requires specific design and pump selection considerations. At sewage treatment plant pumping is also required for removal of grit from grit chamber and pumping may be required for conveying separated grease and floating solids to disposal facility.

Generally pumping station should contain at least three pumping units of such capacity to handle the maximum sewage flow if the largest unit is out of service. The pumps should be selected to provide as uniform a flow as possible to the treatment plant. All pumping stations should have an alarm system to signal power or pump failure and every effort should be made to prevent or minimize overflow. Flow measuring device such as venturimeter shall be provided at the pumping station. In all cases raw-sewage pumps should be protected by screens or racks unless special devices such as self cutting grinder pumps are provided. Housing for electric motors should be made above ground and in dry wells electric motors should be provided protection against flooding. Good ventilation in dry well should be provided, preferably of forced air type, and accessibility for repairs and replacements should be ensured.

The site selection for the pumping station is important and the area selected should never get flooded. The station should be easily accessible in all weathers. The stormwater pumping station should be so located that the water may be impounded without causing damage to the properties. Location of the pumping station should be finalize considering the future expansion and expected increase in the sewage flow. There need to be enough space in the pumping station to replace low capacity pump with higher capacities as per the need in future. The capacity of the pumping station is based on the present and future sewage flow. Generally design period up to 15 years is considered for pumps. The civil structure and the pipelines shall be adequate to serve for the design period of 30 years.

Pumping stations are facilities including pumps and equipment for pumping fluids from one place to another. They are used for a variety of infrastructure systems, such as the supply of water to canals, the drainage of low-lying land, and the removal of sewage to processing sites.

A pumping station is, by definition, an integral part of a pumped-storage hydroelectricity installation.

Pumping station Van Sasse in Grave, the Netherlands.

Pumping station Van Sasse in Grave, the Netherlands.

Canal Water Supply

In countries with canal systems, pumping stations are also frequent. Because of the way the system of canal locks work, water is lost from the upper part of a canal each time a vessel passes through. Also, most lock gates are not watertight, so some water leaks from the higher levels of the canal to those lower down. Obviously, the water has to be replaced or eventually the upper levels of the canal would not hold enough water to be navigable.

Canals are usually fed by diverting water from streams and rivers into the upper parts of the canal, but if no suitable source is available, a pumping station can be used to maintain the water level. An excellent example of a canal pumping station is the Claverton Pumping Station on the Kennet and Avon Canal in southern England, United Kingdom. This pumps water from the nearby River Avon to the canal using pumps driven by a waterwheel which is powered by the river.

Where no external water supply is available, back pumping systems may be employed. Water is extracted from the canal below the lowest lock of a flight and is pumped back to the top of the flight, ready for the next boat to pass through. Such installations are usually small.

Land Drainage

New Orleans, Louisiana: Metairie Pumping Station, also known as Pumping Station 6, building, constructed in 1899, near Metairie Road and the head of the 17th Street Canal. Now housing 15 Wood Screw Pumps, it can move over 6 billion US gallons (23,000,000 m³) of water a day.

When low-lying areas of land are drained, the general method is to dig drainage ditches. However, if the area is below sea level then it is necessary to pump the water upwards into water channels that finally drain into the sea.

The Victorians understood this concept, and in the United Kingdom they built pumping stations with water pumps, powered by steam engines to accomplish this task. In Lincolnshire, large areas of wetland at sea level, called The Fens, were turned into rich arable farmland by this method. The land is full of nutrients because of the accumulation of sedimentary mud that created the land initially.

A land drainage pumping station in Sète, France.

Elsewhere, pumping stations are used to remove water that has found its way into low-lying areas as a result of leakage or flooding (in New Orleans, for example).

Package Pumping Station

In more recent times, a "package pumping station" provides an efficient and economic way of installing a drainage system. They are suitable for mechanical building services collection and pumping of liquids like surface water, wastewater or sewage from areas where drainage by gravity is not possible.

A package pumping station is an integrated system, built in a housing manufactured from strong, impact-resistant materials such as precast concrete, polyethylene, or glass-reinforced plastic. The unit is supplied with internal pipework fitted, pre-assembled ready for installation into the ground, after which the submersible pumps and control equipment are fitted. Features may include controls for fully automatic operation; a high-level alarm indication, in the event of pump failure; and possibly a guide-rail/auto-coupling/pedestal system, to permit easy removal of pumps for maintenance.

Traditional site constructed systems have the valve vault components installed in a separate structure. Having two structural components can lead to potentially serious site problems such as uneven settling between components which results in stress on, and failure of the pipes and connections between components. The development of a packaged pump station system combined all components into a single housing which not only eliminates uneven settling issues, but pre-plumbing and outfitting each unit prior to installation can reduce the cost and time involved with civil work and site labor.

Sewage Systems

The now flooded C Station at Abbey Mills in London. The concrete platforms used to house large motor / pump assemblies that brought sewage up from a deep main drain into several outfall sewers, taking it away from the city centre.

Pumping stations in sewage collection systems, also called *lift stations*, are normally designed to handle raw sewage that is fed from underground gravity pipelines (pipes that are sloped so that a liquid can flow in one direction under gravity). Sewage is fed into and stored in an underground pit, commonly known as a *wet well*. The well is equipped with electrical instrumentation to detect the level of sewage present. When the sewage level rises to a predetermined point, a pump will be started to lift the sewage upward through a pressurized pipe system called a *sewer force main* or rising main from where the sewage is discharged into a gravity manhole. From here the cycle starts all over again until the sewage reaches its point of destination—usually a treatment plant. By this method, pumping stations are used to move waste to higher elevations. In the case of high sewage flows into the well (for example during peak flow periods and wet weather) additional pumps will be used. If this is insufficient, or in the case of failure of the pumping station, a backup in the sewer system can occur, leading to a sanitary sewer overflow—the discharge of raw sewage into the environment.

Sewage pumping stations are typically designed so that one pump or one set of pumps will handle normal peak flow conditions. Redundancy is built into the system so that in the event that any one pump is out of service, the remaining pump or pumps will handle the designed flow. The storage volume of the wet well between the "pump on" and "pump off" settings is designed to minimize pump starts and stops, but is not so long a retention time as to allow the sewage in the wet well to go septic.

Sewage pumps are almost always end-suction centrifugal pumps with open impellers and are specially designed with a large open passage so as to avoid clogging with debris or winding stringy debris onto the impeller. A four pole or six pole AC induction motor normally drives the pump. Rather than provide large open passages, some pumps, typically smaller sewage pumps, also macerate any solids within the sewage breaking them down into smaller parts which can more easily pass through the impeller.

The interior of a sewage pump station is a very dangerous place. Poisonous gases, such as methane and hydrogen sulfide, can accumulate in the wet well; an ill-equipped person entering the well would be overcome by fumes very quickly. Any entry into the wet well requires the correct confined space entry method for a hazardous environment. To minimize the need for entry, the facility is normally designed to allow pumps and other equipment to be removed from outside the wet well.

Traditional sewage pumping stations incorporate both a wet well and a "dry well". Often these are the same structure separated by an internal divide. In this configuration pumps are installed below ground level on the base of the dry well so that their inlets are below water level on pump start, priming the pump and also maximising the available NPSH. Although nominally isolated from the sewage in the wet well, dry wells are underground, confined spaces and require appropriate precautions for entry. Further, any failure or leakage of the pumps or pipework can discharge sewage directly into the dry well with complete flooding not an uncommon occurrence. As a result, the electric motors are normally mounted above the overflow, top water level of the wet well, usually above ground level, and drive the sewage pumps through an extended vertical shaft. To protect the above ground motors from weather, small pump houses are normally built, which also incorporate the electrical switchgear and control electronics. These are the visible parts of a traditional sewage pumping station although they are typically smaller than the underground wet and dry wells.

More modern pumping stations do not require a dry well or pump house and usually consist only of a wet well. In this configuration, submersible sewage pumps with closely coupled electric motor are mounted within the wet well itself, submerged within the sewage. Submersible pumps are mounted on two vertical guide rails and seal onto a permanently fixed "duckfoot", which forms both a mount and also a vertical bend for the discharge pipe. For maintenance or replacement, submersible pumps are raised by a chain off of the duckfoot and up the two guide rails to the maintenance (normally ground) level. Reinstalling the pumps simply reverses this process with the pump being remounted on the guide rails and lowered onto the duckfoot where the weight of the pump reseals it. As the motors are sealed and weather is not a concern, no above ground structures are required, excepting a small kiosk to contain the electrical switchgear and control systems.

Due to the much reduced health and safety concerns, and smaller footprint and visibility, submersible pump sewage pumping stations have almost completely superseded traditional sewage pumping stations. Further, a refit of a traditional pumping station usually involves converting it into a modern pumping station by installing submersibles in the wet well, demolishing the pump house and retiring the dry well by either stripping it, or knocking down the internal partition and merging it with the wet well.

Electronic Controllers

The pump manufacturers have always designed and manufactured electronic devices to control and supervise the pumping stations. Today it is also very common to use a programmable logic controller (PLC) or Remote Terminal Unit (RTU) to do such work, but the experience needed to solve certain particular problems, makes an easy choice to look for a specific pump controller. RTUs are very helpful in remote monitoring of each pumping station from a centralized control room with SCADA (Supervisory Control & Data Acquisition) systems. This setup can be helpful in monitoring pump faults, levels or any other alarms and parameters making it more efficient.

Pumped-storage Schemes

A pumped-storage scheme is a type of power station for storing and producing electricity to supply high peak demands by moving water between reservoirs at different elevations.

Typically, water is channeled from a high-level reservoir to a low-level reservoir, through turbine generators that generate electricity. This is done when the station is required to generate power. During low-demand periods, such as overnight, the generators are reversed to become pumps that move the water back up to the top reservoir.

List of Pumping Stations

There are countless thousands of pumping stations throughout the world. The following is a list of those described in this encyclopaedia.

United Kingdom

In the UK, during the Victorian Era, there was a fashion for public buildings to feature highly ornate architecture. Consequently, a considerable number of former pumping stations have been listed and preserved. The majority were originally steam-powered, and where the steam engines are still *in situ*, many of the sites have since re-opened as museum attractions.

Canal Water Supply

- Claverton Pumping Station, on the Kennet and Avon Canal, near Bath, water-powered
- Cobb's Engine House, ruin near southern portal of Netherton Tunnel
- Crofton Pumping Station, on the Kennet and Avon Canal, near Great Bedwyn
- Leawood Pump House, on the Cromford Canal in Derbyshire

- Smethwick Engine, now removed from original site to Birmingham Thinktank
- New Smethwick Pumping Station (now part of Galton Valley Canal Heritage Centre)

Groundwater Supply

Used to pump water from a well into a reservoir

- Bestwood Pumping Station, Nottinghamshire
- Boughton Pumping Station, Nottinghamshire
- Bratch Pumping Station, Staffordshire
- Mill Meece Pumping Station, in Staffordshire
- Papplewick Pumping Station, Nottinghamshire *(pumped from a 200 ft (61 m) deep well)*
- Selly Oak Pumping Station, Birmingham *(building converted to an electricity sub-station)*

Hydraulic Power Station

- Wapping Hydraulic Power Station, London *(converted to electricity, now an arts centre and restaurant)*

Land Drainage

- Pinchbeck Engine, near Spalding *(preserved beam engine and scoop wheel)*
- Pode Hole pumping station, near Spalding, Lincolnshire *(formerly steam beam engines, no longer present)*
- Prickwillow Engine House, near Ely, Cambridgeshire *(now the Museum of Fenland Drainage)*
- Stretham Old Engine, Stretham, Cambridgeshire
- Westonzoyland Pumping Station, Somerset

Public Water Supply

Used to pump drinking water from a reservoir into a water supply system.

- Blagdon Pumping Station, Chew Valley, Somerset
- Edgbaston Waterworks, Birmingham *(probably not a "museum" site)*
- Kempton Park Pumping Station, London
- Kew Bridge Pumping Station, Kew Bridge, London
- Langford Pumping Station ("Museum of Power"), Essex

- Ryhope Engines Museum, Sunderland
- Tees Cottage Pumping Station, Darlington

Sewage

- Abbey Pumping Station, Leicester
- Abbey Mills Pumping Station, in North London. *(steam engines no longer present)*
- Cheddars Lane Pumping Station, Cambridge
- Claymills Pumping Station, near Burton upon Trent
- Coleham Pumping Station, Coleham, near Shrewsbury
- Crossness Pumping Station, in South London
- Dock Road Edwardian Pumping Station, in Northwich, Cheshire *(Gas engines. Built 1913)*
- Low Hall Pumping Station, Walthamstow, North London
- Markfield Beam Engine, Tottenham, London
- Old Brook Pumping Station, Chatham, Kent

Underground Railway

- Brunel Engine House (now Brunel Museum), Rotherhithe, East London *(extracted water from Thames Tunnel; engine no longer present)*
- Shore Road Pumping Station, Birkenhead, Wirral *(originally steam, now electric; extracts water from the rail tunnel under the River Mersey)*

Hong Kong

Public Water Supply

- Engineer's Office of the Former Pumping Station, Hong Kong

Iraq

Agricultural Drainage

- Nasiriyah Drainage Pump Station, Dhi Qar Province

Netherlands

Land Drainage

- Cruquius pumping station *(Operational, but no longer steam-powered.)*

  – an 8-beam Cornish engine with the largest cylinder (144 in (3.5m) diameter) in the world.

- ir.D.F. Woudagemaal, (ir. Wouda pumping station) *(world's largest steam-powered pumping station)*

Spain

- Stations for public water supply in Barcelona. One of them is a Barcelona City History Museum heritage site *(MUHBA Casa de l'aigua)*. Another is a museum itself: *Museu Agbar de les Aigües* (Agbar water museum).

5

Wastewater Treatment and its Methods

Wastewater treatment is a method to convert wastewater into a product that can be disposed into water bodies so that it causes minimal harm. Wastewater goes through processes such as water treatment, stormwater management and sewage treatment. This chapter discusses the methods of various kinds of treatment methods in a critical manner providing key analysis to the subject matter.

Wastewater Treatment

Wastewater treatment is a process used to convert wastewater - which is water no longer needed or suitable for its most recent use - into an effluent that can be either returned to the water cycle with minimal environmental issues or reused. The latter is called water reclamation and implies avoidance of disposal by use of treated wastewater effluent for various purposes. Treatment means removing impurities from water being treated; and some methods of treatment are applicable to both water and wastewater. The physical infrastructure used for wastewater treatment is called a wastewater treatment plant (WWTP).

The treatment of wastewater belongs to the overarching field of Public Works - Environmental, with the management of human waste, solid waste, sewage treatment, stormwater (drainage) management, and water treatment. By-products from wastewater treatment plants, such as screenings, grit and sewage sludge may also be treated in a wastewater treatment plant. If the wastewater is predominantly from municipal sources (households and small industries) it is called sewage and its treatment is called sewage treatment.

Disposal or Reuse

Although disposal or reuse occurs after treatment, it must be considered first. Since disposal or reuse are the objectives of wastewater treatment, disposal or reuse options are the basis for treatment decisions. Acceptable impurity concentrations may vary with the type of use or location of disposal. Transportation costs often make acceptable impurity concentrations dependent upon location of disposal, but expensive treatment requirements may encourage selection of a disposal location on the basis of impurity concentrations. Ocean disposal is subject to international treaty requirements. International treaties may also regulate disposal into rivers crossing international borders. Water bodies entirely within the jurisdiction of a single nation may be subject to regulations of multiple local governments. Acceptable impurity concentrations may vary widely among different jurisdictions for disposal of wastewater to evaporation ponds, infiltration basins, or injection wells.

Processes Used

Phase Separation

Clarifiers are widely used for wastewater treatment.

Phase separation transfers impurities into a non-aqueous phase. Phase separation may occur at intermediate points in a treatment sequence to remove solids generated during oxidation or polishing. Grease and oil may be recovered for fuel or saponification. Solids often require dewatering of sludge in a wastewater treatment plant. Disposal options for dried solids vary with the type and concentration of impurities removed from water.

Production of waste brine, however, may discourage wastewater treatment removing dissolved inorganic solids from water by methods like ion exchange, reverse osmosis, and distillation.

Sedimentation

Solids and non-polar liquids may be removed from wastewater by gravity when density differences are sufficient to overcome dispersion by turbulence. Gravity separation of solids is the primary treatment of sewage, where the unit process is called "primary settling tanks" or "primary sedimentation tanks". It is also widely used for the treatment of other wastewaters. Solids that are heavier than water will accumulate at the bottom of quiescent settling basins. More complex clarifiers also have skimmers to simultaneously remove floating grease like soap scum and solids like feathers or wood chips. Containers like the API oil-water separator are specifically designed to separate non-polar liquids.

Primary settling tank of wastewater treatment plant in Dresden-Kaditz, Germany

Filtration

Colloidal suspensions of fine solids may be removed by filtration through fine physical barriers distinguished from coarser screens or sieves by the ability to remove particles smaller than the openings through which the water passes. Other types of water filters remove impurities by chemical or biological processes described below.

Oxidation

Oxidation reduces the biochemical oxygen demand of wastewater, and may reduce the toxicity of some impurities. Secondary treatment converts some impurities to carbon dioxide, water, and biosolids. Chemical oxidation is widely used for disinfection.

Aeration tank of an activated sludge process at the wastewater treatment plant in Dresden-Kaditz, Germany

Biochemical Oxidation

Secondary treatment by biochemical oxidation of dissolved and colloidal organic compounds is widely used in sewage treatment and is applicable to some agricultural and industrial waste waters. Biological oxidation will preferentially remove organic compounds useful as a food supply for the treatment ecosystem. Concentration of some less digestible compounds may be reduced by co-metabolism. Removal efficiency is limited by the minimum food concentration required to sustain the treatment ecosystem. Removal efficiency of over 96% are readily attainable through proper process controls and effective maintenance of plant equipment.

Chemical Oxidation

Chemical oxidation may remove some persistent organic pollutants and concentrations remaining after biochemical oxidation. Disinfection by chemical oxidation kills bacteria and microbial pathogens by adding ozone, chlorine or hypochlorite to wastewater.

Polishing

Polishing refers to treatments made following the above methods. These treatments may also be used independently for some industrial wastewater. Chemical reduction or pH adjustment minimizes chemical reactivity of wastewater following chemical oxidation. Carbon filtering removes remaining contaminants and impurities by chemical absorption onto activated carbon. Filtration through sand (calcium carbonate) or fabric filters is the most common method used in municipal wastewater treatment.

Wastewater Treatment Plants

Wastewater treatment plants may be distinguished by the type of wastewater to be treated, i.e. whether it is sewage, industrial wastewater, agricultural wastewater or leachate.

Overview of the wastewater treatment plant of Antwerpen-Zuid, located in the south of the agglomeration of Antwerp (Belgium)

Sewage Treatment Plants

A typical municipal sewage treatment plant in an industrialized country may include primary treatment to remove solid material, secondary treatment to digest dissolved and suspended organic material as well as the nutrients nitrogen and phosphorus, and - sometimes but not always - disinfection to kill pathogenic bacteria. The sewage sludge that is produced in sewage treatment plants undergoes sludge treatment. Larger municipalities often include factories discharging industrial wastewater into the municipal sewer system. The term "sewage treatment plant" is now often replaced with the term "wastewater treatment plant".

Tertiary Treatment

Tertiary treatment is a term applied to polishing methods used following a traditional sewage treatment sequence. Tertiary treatment is being increasingly applied in industrialized countries and most common technologies are micro filtration or synthetic membranes. After membrane filtration, the treated wastewater is nearly indistinguishable from waters of natural origin of drinking quality (without its minerals). Nitrates can be removed from wastewater by natural processes in wetlands but also via microbial denitrification. Ozone wastewater treatment is also growing in popularity, and requires the use of an ozone generator, which decontaminates the water as ozone bubbles percolate through the tank, but this treatment is energy intensive. Latest, and very promising treatment technology is the use aerobic granulation.

Industrial Wastewater Treatment Plants

Disposal of wastewaters from an industrial plant is a difficult and costly problem. Most petroleum refineries, chemical and petrochemical plants have onsite facilities to treat their wastewaters so that the pollutant concentrations in the treated wastewater comply with the local and/or national regulations regarding disposal of wastewaters into community treatment plants or into rivers, lakes or oceans. Constructed wetlands are being used in an increasing number of cases as they provided high quality and productive on-site treatment. Other industrial processes that produce a lot of waste-waters such as paper and pulp production has created environmental concern, leading to development of processes to recycle water use within plants before they have to be cleaned and disposed.

Industrial wastewater treatment plants are required where municipal sewage treatment plants are unavailable or cannot adequately treat specific industrial wastewaters. Industrial wastewater plants may reduce raw water costs by converting selected wastewaters to reclaimed water used for different purposes. Industrial wastewater treatment plants may reduce wastewater treatment charges collected by municipal sewage treatment plants by pre-treating wastewaters to reduce concentrations of pollutants measured to determine user fees.

Although economies of scale may favor use of a large municipal sewage treatment plant for disposal of small volumes of industrial wastewater, industrial wastewater treatment and disposal may be less expensive than correctly apportioned costs for larger volumes of industrial wastewater not requiring the conventional sewage treatment sequence of a small municipal sewage treatment plant.

An industrial wastewater treatment plant may include one or more of the following rather than the conventional primary, secondary, and disinfection sequence of sewage treatment:

- An API oil-water separator, for removing separate phase oil from wastewater.
- A clarifier, for removing solids from wastewater.
- A roughing filter, to reduce the biochemical oxygen demand of wastewater.
- A carbon filtration plant, to remove toxic dissolved organic compounds from wastewater.
- An advanced electrodialysis reversal (EDR) system with ion exchange membranes.

Agricultural Wastewater Treatment Plants

Agricultural wastewater treatment for continuous confined animal operations like milk and egg production may be performed in plants using mechanized treatment units similar to those described under industrial wastewater; but where land is available for ponds, settling basins and facultative lagoons may have lower operational costs for seasonal use conditions from breeding or harvest cycles.

Leachate Treatment Plants

Leachate treatment plants are used to treat leachate from landfills. Treatment options include: biological treatment, mechanical treatment by ultrafiltration, treatment with active carbon filters and reverse osmosis using disc tube module technology.

Industrial Wastewater Treatment

Industrial wastewater treatment covers the mechanisms and processes used to treat wastewater that is produced as a by-product of industrial or commercial activities. After treatment, the treated industrial wastewater (or effluent) may be reused or released to a sanitary sewer or to a surface water in the environment. Most industries produce some wastewater although recent trends in the developed world have been to minimise such production or recycle such wastewater within the production process. However, many industries remain dependent on processes that produce wastewaters.

Sources Of Industrial Wastewater

Complex Organic Chemicals Industry

A range of industries manufacture or use complex organic chemicals. These include pesticides, pharmaceuticals, paints and dyes, petrochemicals, detergents, plastics, paper pollution, etc. Waste waters can be contaminated by feedstock materials, by-products, product material in soluble or particulate form, washing and cleaning agents, solvents and added value products such as plasticisers. Treatment facilities that do not need control of their effluent typically opt for a type of aerobic treatment, i.e. aerated lagoons.

Electric Power Plants

Fossil-fuel power stations, particularly coal-fired plants, are a major source of industrial wastewater. Many of these plants discharge wastewater with significant levels of metals such as lead, mercury, cadmium and chromium, as well as arsenic, selenium and nitrogen compounds (nitrates and nitrites). Wastewater streams include flue-gas desulfurization, fly ash, bottom ash and flue gas mercury control. Plants with air pollution controls such as wet scrubbers typically transfer the captured pollutants to the wastewater stream.

Ash ponds, a type of surface impoundment, are a widely used treatment technology at coal-fired plants. These ponds use gravity to settle out large particulates (measured as total suspended solids) from power plant wastewater. This technology does not treat dissolved pollutants. Power stations use additional technologies to control pollutants, depending on the particular wastestream in the plant. These include dry ash handling, closed-loop ash recycling, chemical precipitation, biological treatment (such as an activated sludge process), and evaporation.

Food Industry

Wastewater generated from agricultural and food operations has distinctive characteristics that set it apart from common municipal wastewater managed by public or private sewage treatment plants throughout the world: it is biodegradable and non-toxic, but has high concentrations of biochemical oxygen demand (BOD) and suspended solids (SS). The constituents of food and agriculture wastewater are often complex to predict, due to the differences in BOD and pH in effluents from vegetable, fruit, and meat products and due to the seasonal nature of food processing and post-harvesting.

Processing of food from raw materials requires large volumes of high grade water. Vegetable washing generates waters with high loads of particulate matter and some dissolved organic matter. It may also contain surfactants.

Animal slaughter and processing produces very strong organic waste from body fluids, such as blood, and gut contents. This wastewater is frequently contaminated by significant levels of antibiotics and growth hormones from the animals and by a variety of pesticides used to control external parasites.

Processing food for sale produces wastes generated from cooking which are often rich in plant organic material and may also contain salt, flavourings, colouring material and acids or alkali. Very significant quantities of oil or fats may also be present.

Iron and Steel Industry

The production of iron from its ores involves powerful reduction reactions in blast furnaces. Cooling waters are inevitably contaminated with products especially ammonia and cyanide. Production of coke from coal in coking plants also requires water cooling and the use of water in by-products separation. Contamination of waste streams includes gasification products such as benzene, naphthalene, anthracene, cyanide, ammonia, phenols, cresols together with a range of more complex organic compounds known collectively as polycyclic aromatic hydrocarbons (PAH).

The conversion of iron or steel into sheet, wire or rods requires hot and cold mechanical transformation stages frequently employing water as a lubricant and coolant. Contaminants include hydraulic oils, tallow and particulate solids. Final treatment of iron and steel products before onward sale into manufacturing includes *pickling* in strong mineral acid to remove rust and prepare the surface for tin or chromium plating or for other surface treatments such as galvanisation or painting. The two acids commonly used are hydrochloric acid and sulfuric acid. Wastewaters include acidic rinse waters together with waste acid. Although many plants operate acid recovery plants (particularly those using hydrochloric acid), where the mineral acid is boiled away from the iron salts, there remains a large volume of highly acid ferrous sulfate or ferrous chloride to be disposed of. Many steel industry wastewaters are contaminated by hydraulic oil, also known as *soluble oil*.

Mines and Quarries

Mine wastewater effluent in Peru, with neutralized pH from tailing runoff.

The principal waste-waters associated with mines and quarries are slurries of rock particles in water. These arise from rainfall washing exposed surfaces and haul roads and also from rock washing and grading processes. Volumes of water can be very high, especially rainfall related arisings on large sites. Some specialized separation operations, such as coal washing to separate coal from native rock using density gradients, can produce wastewater contaminated by fine particulate haematite and surfactants. Oils and hydraulic oils are also common contaminants.

Wastewater from metal mines and ore recovery plants are inevitably contaminated by the minerals present in the native rock formations. Following crushing and extraction of the desirable materials, undesirable materials may enter the wastewater stream. For metal mines, this can include unwanted metals such as zinc and other materials such as arsenic. Extraction of high value metals such as gold and silver may generate slimes containing very fine particles in where physical removal of contaminants becomes particularly difficult.

Additionally, the geologic formations that harbour economically valuable metals such as copper and gold very often consist of sulphide-type ores. The processing entails grinding the rock into fine particles and then extracting the desired metal(s), with the leftover rock being known as tailings. These tailings contain a combination of not only undesirable leftover metals, but also sulphide components which eventually form sulphuric acid upon the exposure to air and water that inevitably occurs when the tailings are disposed of in large impoundments. The resulting acid mine drainage, which is often rich in heavy metals (because acids dissolve metals), is one of the many environmental impacts of mining.

Nuclear Industry

The waste production from the nuclear and radio-chemicals industry is dealt with as *Radioactive waste.*

Pulp and Paper Industry

Effluent from the pulp and paper industry is generally high in suspended solids and BOD. Plants that bleach wood pulp for paper making may generate chloroform, dioxins (including 2,3,7,8-TCDD), furans, phenols and chemical oxygen demand (COD). Stand-alone paper mills using imported pulp may only require simple primary treatment, such as sedimentation or dissolved air flotation. Increased BOD or COD loadings, as well as organic pollutants, may require biological treatment such as activated sludge or upflow anaerobic sludge blanket reactors. For mills with high inorganic loadings like salt, tertiary treatments may be required, either general membrane treatments like ultrafiltration or reverse osmosis or treatments to remove specific contaminants, such as nutrients.

Industrial Oil Contamination

Industrial applications where oil enters the wastewater stream may include vehicle wash bays, workshops, fuel storage depots, transport hubs and power generation. Often the wastewater is discharged into local sewer or trade waste systems and must meet local environmental specifications. Typical contaminants can include solvents, detergents, grit. lubricants and hydrocarbons.

Water Treatment

Many industries have a need to treat water to obtain very high quality water for demanding purposes such as environmental discharge compliance. Water treatment produces organic and mineral sludges from filtration and sedimentation. Ion exchange using natural or synthetic resins removes calcium, magnesium and carbonate ions from water, typically replacing them with sodium, chloride, hydroxyl and/or other ions. Regeneration of ion exchange columns with strong acids and alkalis produces a wastewater rich in hardness ions which are readily precipitated out, especially when in admixture with other wastewater constituents.

Wool Processing

Insecticide residues in fleeces are a particular problem in treating waters generated in wool processing. Animal fats may be present in the wastewater, which if not contaminated, can be recovered for the production of tallow or further rendering.

Treatment of Industrial Wastewater

The various types of contamination of wastewater require a variety of strategies to remove the contamination.

Brine Treatment

Brine treatment involves removing dissolved salt ions from the waste stream. Although similarities to seawater or brackish water desalination exist, industrial brine treatment may contain unique combinations of dissolved ions, such as hardness ions or other metals, necessitating specific processes and equipment.

Brine treatment systems are typically optimized to either reduce the volume of the final discharge for more economic disposal (as disposal costs are often based on volume) or maximize the recovery of fresh water or salts. Brine treatment systems may also be optimized to reduce electricity consumption, chemical usage, or physical footprint.

Brine treatment is commonly encountered when treating cooling tower blowdown, produced water from steam assisted gravity drainage (SAGD), produced water from natural gas extraction such as coal seam gas, frac flowback water, acid mine or acid rock drainage, reverse osmosis reject, chlor-alkali wastewater, pulp and paper mill effluent, and waste streams from food and beverage processing.

Brine treatment technologies may include: membrane filtration processes, such as reverse osmosis; ion exchange processes such as electrodialysis or weak acid cation exchange; or evaporation processes, such as brine concentrators and crystallizers employing mechanical vapour recompression and steam.

Reverse osmosis may not be viable for brine treatment, due to the potential for fouling caused by hardness salts or organic contaminants, or damage to the reverse osmosis membranes from hydrocarbons.

Evaporation processes are the most widespread for brine treatment as they enable the highest degree of concentration, as high as solid salt. They also produce the highest purity effluent, even distillate-quality. Evaporation processes are also more tolerant of organics, hydrocarbons, or hardness salts. However, energy consumption is high and corrosion may be an issue as the prime mover is concentrated salt water. As a result, evaporation systems typically employ titanium or duplex stainless steel materials.

Brine Management

Brine management examines the broader context of brine treatment and may include consideration of government policy and regulations, corporate sustainability, environmental impact, recycling, handling and transport, containment, centralized compared to on-site treatment, avoidance and reduction, technologies, and economics. Brine management shares some issues with leachate management and more general waste management.

Solids Removal

Most solids can be removed using simple sedimentation techniques with the solids recovered as slurry or sludge. Very fine solids and solids with densities close to the density of water pose special problems. In such case filtration or ultrafiltration may be required. Although, flocculation may be used, using alum salts or the addition of polyelectrolytes.

Oils and Grease Removal

The effective removal of oils and grease is dependent on the characteristics of the oil in terms of its suspension state and droplet size, which will in turn affect the choice of separator technology.

Oil pollution in water usually comes in four states, often in combination:

- free oil - large oil droplets sitting on the surface;
- heavy oil, which sits at the bottom, often adhering to solids like dirt;
- emulsified, where the oil droplets are heavily "chopped"; and
- dissolved oil, where the droplets are fully dispersed and not visible. Emulsified oil droplets are the most common in industrial oily wastewater and are extremely difficult to separate.

The methodology for separating the oil is dependent on the oil droplet size. Larger oil droplets such as those in free oil pollution are easily removed, but as the droplets become smaller, some separator technologies perform better than others.

Most separator technologies will have an optimum range of oil droplet sizes that can be effectively treated. This is known as the "micron rating."

Analysing the oily water to determine droplet size can be performed with a video particle analyser. Alternatively, there are commonalities in industries for oil droplet sizes. Larger droplets–greater than 60 microns–are often present in wastewater in workshops, re-fuel areas and depots. Twenty to 50 micron oil droplets often are present in vehicle wash bays, meat processing and dairy manufacturing effluent and aluminium billet cooling towers. Smaller droplets in the range of 10 to 20 microns tend to occur in workshops and condensates.

Each separator technology will have its' own performance curve outlining optimum performance based on oil droplet size. the most common separators are gravity tanks or pits, API oil-water separators or plate packs, chemical treatment via DAFs, centrifuges, media filters and hydrocyclones.

API Separators

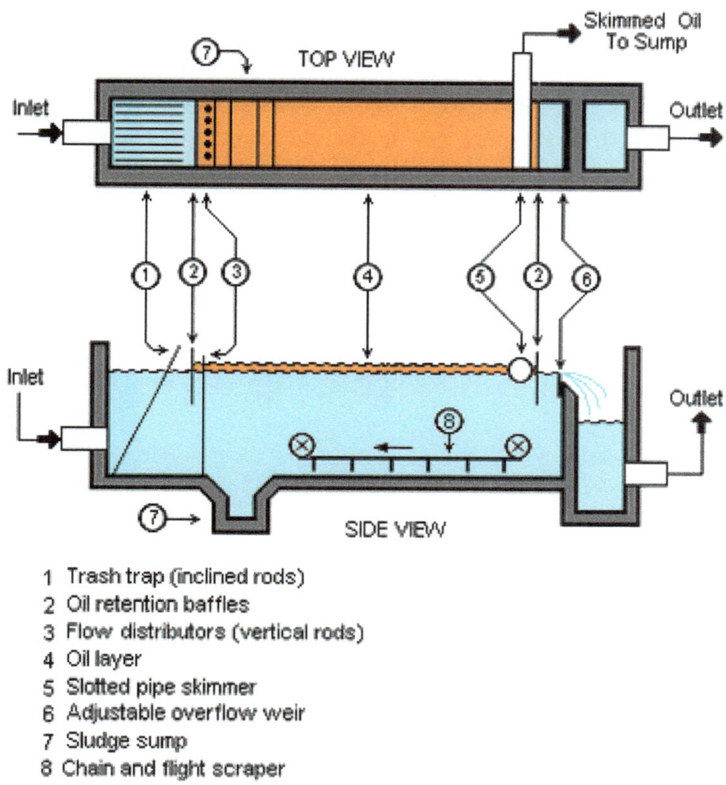

A typical API oil-water separator used in many industries

Many oils can be recovered from open water surfaces by skimming devices. Considered a dependable and cheap way to remove oil, grease and other hydrocarbons from water, oil skimmers can sometimes achieve the desired level of water purity. At other times, skimming is also a cost-efficient method to remove most of the oil before using membrane filters and chemical processes. Skimmers will prevent filters from blinding prematurely and keep chemical costs down because there is less oil to process.

Because grease skimming involves higher viscosity hydrocarbons, skimmers must be equipped with heaters powerful enough to keep grease fluid for discharge. If floating grease forms into solid clumps or mats, a spray bar, aerator or mechanical apparatus can be used to facilitate removal.

However, hydraulic oils and the majority of oils that have degraded to any extent will also have a soluble or emulsified component that will require further treatment to eliminate. Dissolving or emulsifying oil using surfactants or solvents usually exacerbates the problem rather than solving it, producing wastewater that is more difficult to treat.

The wastewaters from large-scale industries such as oil refineries, petrochemical plants, chemical plants, and natural gas processing plants commonly contain gross amounts of oil and suspended solids. Those industries use a device known as an API oil-water separator which is designed to separate the oil and suspended solids from their wastewater effluents. The name is derived from the fact that such separators are designed according to standards published by the American Petroleum Institute (API).

The API separator is a gravity separation device designed by using Stokes Law to define the rise velocity of oil droplets based on their density and size. The design is based on the specific gravity difference between the oil and the wastewater because that difference is much smaller than the specific gravity difference between the suspended solids and water. The suspended solids settles to the bottom of the separator as a sediment layer, the oil rises to top of the separator and the cleansed wastewater is the middle layer between the oil layer and the solids.

Typically, the oil layer is skimmed off and subsequently re-processed or disposed of, and the bottom sediment layer is removed by a chain and flight scraper (or similar device) and a sludge pump. The water layer is sent to further treatment for additional removal of any residual oil and then to some type of biological treatment unit for removal of undesirable dissolved chemical compounds.

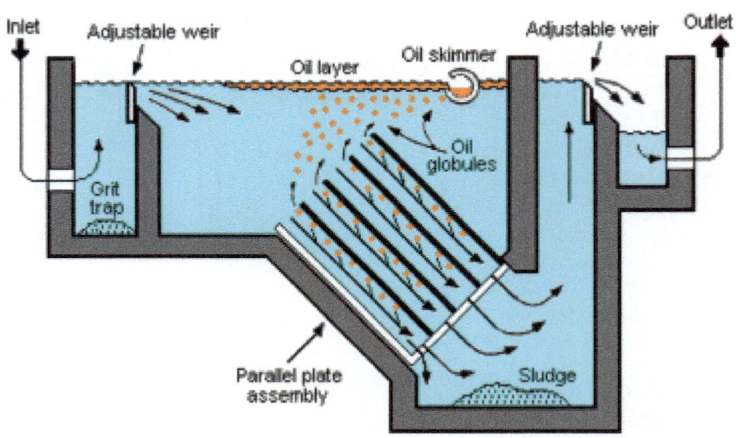

A typical parallel plate separator

Parallel plate separators are similar to API separators but they include tilted parallel plate assemblies (also known as parallel packs). The parallel plates provide more surface for suspended oil droplets to coalesce into larger globules. Such separators still depend upon the specific gravity between the suspended oil and the water. However, the parallel plates enhance the degree of oil-water separation. The result is that a parallel plate separator requires significantly less space than a conventional API separator to achieve the same degree of separation.

Hydrocyclone Oil Separators

Hydrocyclone oil separators operate on the process where wastewater enters the cyclone chamber and is spun under extreme centrifugal forces more than 1000 times the force of gravity. This force causes the water and oil droplets to separate. The separated oil is discharged from one end of the cyclone where treated water is discharged through the opposite end for further treatment, filtration or discharge.

Hydrocyclones are useful for the greatest range of oil droplet sizes operating from less than 10 microns and up and can operate continuously without water pre-treatment and at any temperature and pH. Applications where hydrocyclones are found are in industry where oily water sources arise in workshops, vehicle wash bays, transport hubs, fuel depots and aluminium billet processing. Animal fats from meat processing and dairy manufacturing can also be removed without the need of chemical treatment that often is required for dissolved air flotation (DAF) systems.

Removal of Biodegradable Organics

Biodegradable organic material of plant or animal origin is usually possible to treat using extended conventional sewage treatment processes such as activated sludge or trickling filter. Problems can arise if the wastewater is excessively diluted with washing water or is highly concentrated such as undiluted blood or milk. The presence of cleaning agents, disinfectants, pesticides, or antibiotics can have detrimental impacts on treatment processes.

Activated Sludge Process

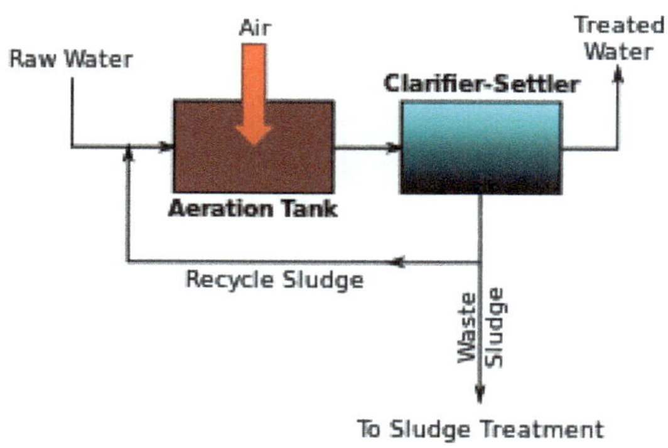

A generalized diagram of an activated sludge process.

Activated sludge is a biochemical process for treating sewage and industrial wastewater that uses air (or oxygen) and microorganisms to biologically oxidize organic pollutants, producing a waste sludge (or floc) containing the oxidized material. In general, an activated sludge process includes:

- An aeration tank where air (or oxygen) is injected and thoroughly mixed into the wastewater.
- A settling tank (usually referred to as a clarifier or "settler") to allow the waste sludge to settle. Part of the waste sludge is recycled to the aeration tank and the remaining waste sludge is removed for further treatment and ultimate disposal.

Trickling Filter Process

A trickling filter consists of a bed of rocks, gravel, slag, peat moss, or plastic media over which wastewater flows downward and contacts a layer (or film) of microbial slime covering the bed media. Aerobic conditions are maintained by forced air flowing through the bed or by natural convection of air. The process involves adsorption of organic compounds in the wastewater by the microbial slime layer, diffusion of air into the slime layer to provide the oxygen required for the biochemical oxidation of the organic compounds. The end products include carbon dioxide gas, water and other products of the oxidation. As the slime layer thickens, it becomes difficult for the air to penetrate the layer and an inner anaerobic layer is formed.

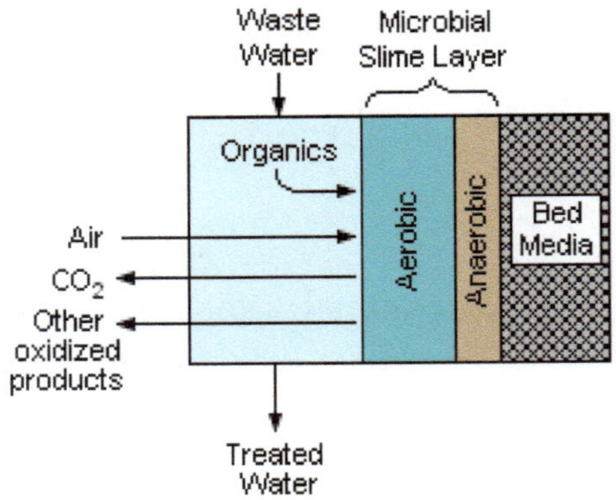

A schematic cross-section of the contact face of the bed media in a trickling filter

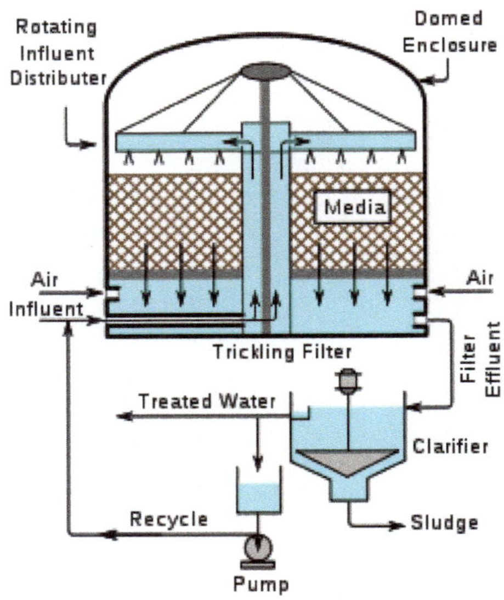

A typical complete trickling filter system

The fundamental components of a complete trickling filter system are:

- A bed of filter medium upon which a layer of microbial slime is promoted and developed.
- An enclosure or a container which houses the bed of filter medium.
- A system for distributing the flow of wastewater over the filter medium.
- A system for removing and disposing of any sludge from the treated effluent.

The treatment of sewage or other wastewater with trickling filters is among the oldest and most well characterized treatment technologies.

A trickling filter is also often called a *trickle filter*, *trickling biofilter*, *biofilter*, *biological filter* or *biological trickling filter*.

Treatment of other Organics

Synthetic organic materials including solvents, paints, pharmaceuticals, pesticides, products from coke production and so forth can be very difficult to treat. Treatment methods are often specific to the material being treated. Methods include advanced oxidation processing, distillation, adsorption, vitrification, incineration, chemical immobilisation or landfill disposal. Some materials such as some detergents may be capable of biological degradation and in such cases, a modified form of wastewater treatment can be used.

Treatment of Acids and Alkalis

Acids and alkalis can usually be neutralised under controlled conditions. Neutralisation frequently produces a precipitate that will require treatment as a solid residue that may also be toxic. In some cases, gases may be evolved requiring treatment for the gas stream. Some other forms of treatment are usually required following neutralisation.

Waste streams rich in hardness ions as from de-ionisation processes can readily lose the hardness ions in a buildup of precipitated calcium and magnesium salts. This precipitation process can cause severe *furring* of pipes and can, in extreme cases, cause the blockage of disposal pipes. A 1 metre diameter industrial marine discharge pipe serving a major chemicals complex was blocked by such salts in the 1970s. Treatment is by concentration of de-ionisation waste waters and disposal to landfill or by careful pH management of the released wastewater.

Treatment of Toxic Materials

Toxic materials including many organic materials, metals (such as zinc, silver, cadmium, thallium, etc.) acids, alkalis, non-metallic elements (such as arsenic or selenium) are generally resistant to biological processes unless very dilute. Metals can often be precipitated out by changing the pH or by treatment with other chemicals. Many, however, are resistant to treatment or mitigation and may require concentration followed by landfilling or recycling. Dissolved organics can be *incinerated* within the wastewater by the advanced oxidation process.

Agricultural Wastewater Treatment

Riparian buffer lining a creek in Iowa

Agricultural wastewater treatment is a farm management agenda for controlling pollution from surface runoff that may be contaminated by chemicals in fertiliser, pesticides, animal slurry, crop residues or irrigation water.

Nonpoint Source Pollution

Nonpoint source pollution from farms is caused by surface runoff from fields during rain storms. Agricultural runoff is a major source of pollution, in some cases the only source, in many watersheds.

Sediment Runoff

Highly erodible soils on a farm in Iowa

Soil washed off fields is the largest source of agricultural pollution in the United States. Excess sediment causes high levels of turbidity in water bodies, which can inhibit growth of aquatic plants, clog fish gills and smother animal larvae.

Farmers may utilize erosion controls to reduce runoff flows and retain soil on their fields. Common techniques include:

- contour ploughing
- crop mulching
- crop rotation
- planting perennial crops
- installing riparian buffers.

Nutrient Runoff

Nitrogen and phosphorus are key pollutants found in runoff, and they are applied to farmland in several ways, such as in the form of commercial fertilizer, animal manure, or municipal or industrial wastewater (effluent) or sludge. These chemicals may also enter runoff from crop residues, irrigation water, wildlife, and atmospheric deposition.

Manure spreader

Farmers can develop and implement nutrient management plans to mitigate impacts on water quality by:

- mapping and documenting fields, crop types, soil types, water bodies
- developing realistic crop yield projections
- conducting soil tests and nutrient analyses of manures and/or sludges applied
- identifying other significant nutrient sources (e.g., irrigation water)
- evaluating significant field features such as highly erodible soils, subsurface drains, and shallow aquifers
- applying fertilizers, manures, and/or sludges based on realistic yield goals and using precision agriculture techniques.

Pesticides

Aerial application (crop dusting) of pesticides over a soybean field in the U.S.

Pesticides are widely used by farmers to control plant pests and enhance production, but chemical pesticides can also cause water quality problems. Pesticides may appear in surface water due to:

- direct application (e.g. aerial spraying or broadcasting over water bodies)
- runoff during rain storms
- aerial drift (from adjacent fields).

Some pesticides have also been detected in groundwater.

Farmers may use Integrated Pest Management (IPM) techniques (which can include biological pest control) to maintain control over pests, reduce reliance on chemical pesticides, and protect water quality.

There are few safe ways of disposing of pesticide surpluses other than through containment in well managed landfills or by incineration. In some parts of the world, spraying on land is a permitted method of disposal.

Point Source Pollution

Farms with large livestock and poultry operations, such as factory farms, can be a major source of point source wastewater. In the United States, these facilities are called *concentrated animal feeding operations* or *confined animal feeding operations* and are being subject to increasing government regulation.

Animal Wastes

Confined Animal Feeding Operation in the United States

The constituents of animal wastewater typically contain

- Strong organic content — much stronger than human sewage
- High solids concentration
- High nitrate and phosphorus content
- Antibiotics

- Synthetic hormones
- Often high concentrations of parasites and their eggs
- Spores of *Cryptosporidium* (a protozoan) resistant to drinking water treatment processes
- Spores of *Giardia*
- Human pathogenic bacteria such as *Brucella* and *Salmonella*

Animal wastes from cattle can be produced as solid or semisolid manure or as a liquid slurry. The production of slurry is especially common in housed dairy cattle.

Treatment

Whilst solid manure heaps outdoors can give rise to polluting wastewaters from runoff, this type of waste is usually relatively easy to treat by containment and/or covering of the heap.

Animal slurries require special handling and are usually treated by containment in lagoons before disposal by spray or trickle application to grassland. Constructed wetlands are sometimes used to facilitate treatment of animal wastes, as are anaerobic lagoons. Excessive application or application to sodden land or insufficient land area can result in direct runoff to watercourses, with the potential for causing severe pollution. Application of slurries to land overlying aquifers can result in direct contamination or, more commonly, elevation of nitrogen levels as nitrite or nitrate.

The disposal of any wastewater containing animal waste upstream of a drinking water intake can pose serious health problems to those drinking the water because of the highly resistant spores present in many animals that are capable of causing disabling disease in humans. This risk exists even for very low-level seepage via shallow surface drains or from rainfall run-off.

Some animal slurries are treated by mixing with straws and composted at high temperature to produce a bacteriologically sterile and friable manure for soil improvement.

Piggery Waste

Hog confinement barn or piggery

Piggery waste is comparable to other animal wastes and is processed as for general animal waste, except that many piggery wastes contain elevated levels of copper that can be toxic in the natural environment. The liquid fraction of the waste is frequently separated off and re-used in the piggery to avoid the prohibitively expensive costs of disposing of copper-rich liquid. Ascarid worms and their eggs are also common in piggery waste and can infect humans if wastewater treatment is ineffective.

Silage Liquor

Fresh or wilted grass or other green crops can be made into a semi-fermented product called silage which can be stored and used as winter forage for cattle and sheep. The production of silage often involves the use of an acid conditioner such as sulfuric acid or formic acid. The process of silage making frequently produces a yellow-brown strongly smelling liquid which is very rich in simple sugars, alcohol, short-chain organic acids and silage conditioner. This liquor is one of the most polluting organic substances known. The volume of silage liquor produced is generally in proportion to the moisture content of the ensiled material.

Treatment

Silage liquor is best treated through prevention by wilting crops well before silage making. Any silage liquor that is produced can be used as part of the food for pigs. The most effective treatment is by containment in a slurry lagoon and by subsequent spreading on land following substantial dilution with slurry. Containment of silage liquor on its own can cause structural problems in concrete pits because of the acidic nature of silage liquor.

Milking Parlour (Dairy Farming) Wastes

Although milk has a deserved reputation as an important and valuable food product, its presence in wastewaters is highly polluting because of its organic strength, which can lead to very rapid de-oxygenation of receiving waters. Milking parlour wastes also contain large volumes of wash-down water, some animal waste together with cleaning and disinfection chemicals.

Treatment

Milking parlour wastes are often treated in admixture with human sewage in a local sewage treatment plant. This ensures that disinfectants and cleaning agents are sufficiently diluted and amenable to treatment. Running milking wastewaters into a farm slurry lagoon is a possible option although this tends to consume lagoon capacity very quickly. Land spreading is also a treatment option.

Slaughtering Waste

Wastewater from slaughtering activities is similar to milking parlour waste although considerably stronger in its organic composition and therefore potentially much more polluting.

Treatment

As for milking parlour waste.

Vegetable Washing Water

Washing of vegetables produces large volumes of water contaminated by soil and vegetable pieces. Low levels of pesticides used to treat the vegetables may also be present together with moderate levels of disinfectants such as chlorine.

Treatment

Most vegetable washing waters are extensively recycled with the solids removed by settlement and filtration. The recovered soil can be returned to the land.

Firewater

Although few farms plan for fires, fires are nevertheless more common on farms than on many other industrial premises. Stores of pesticides, herbicides, fuel oil for farm machinery and fertilizers can all help promote fire and can all be present in environmentally lethal quantities in firewater from fire fighting at farms.

Treatment

All farm environmental management plans should allow for containment of substantial quantities of firewater and for its subsequent recovery and disposal by specialist disposal companies. The concentration and mixture of contaminants in firewater make them unsuited to any treatment method available on the farm. Even land spreading has produced severe taste and odour problems for downstream water supply companies in the past.

Sewage Treatment

Wastewater treatment plant in Massachusetts, United States

Sewage treatment is the process of removing contaminants from wastewater, primarily from household sewage. It includes physical, chemical, and biological processes to remove these contaminants and produce environmentally safe treated wastewater (or treated effluent). A by-product of sewage treatment is usually a semi-solid waste or slurry, called sewage sludge, that has to undergo further treatment before being suitable for disposal or land application.

Sewage treatment may also be referred to as wastewater treatment, although the latter is a broader term which can also be applied to purely industrial wastewater. For most cities, the sewer system will also carry a proportion of industrial effluent to the sewage treatment plant which has usually received pretreatment at the factories themselves to reduce the pollutant load. If the sewer system is a combined sewer then it will also carry urban runoff (stormwater) to the sewage treatment plant.

Terminology

The term "sewage treatment plant" (or "sewage treatment works" in some countries) is nowadays often replaced with the term "wastewater treatment plant".

Sewage can be treated close to where the sewage is created, which may be called a "decentralized" system or even an "on-site" system (in septic tanks, biofilters or aerobic treatment systems). Alternatively, sewage can be collected and transported by a network of pipes and pump stations to a municipal treatment plant. This is called a "centralized" system.

Origins of Sewage

Sewage is generated by residential, institutional, commercial and industrial establishments. It includes household waste liquid from toilets, baths, showers, kitchens, and sinks draining into sewers. In many areas, sewage also includes liquid waste from industry and commerce. The separation and draining of household waste into greywater and blackwater is becoming more common in the developed world, with treated greywater being permitted to be used for watering plants or recycled for flushing toilets.

Sewage Mixing with Rainwater

Sewage may include stormwater runoff or urban runoff. Sewerage systems capable of handling storm water are known as combined sewer systems. This design was common when urban sewerage systems were first developed, in the late 19th and early 20th centuries.[:119] Combined sewers require much larger and more expensive treatment facilities than sanitary sewers. Heavy volumes of storm runoff may overwhelm the sewage treatment system, causing a spill or overflow. Sanitary sewers are typically much smaller than combined sewers, and they are not designed to transport stormwater. Backups of raw sewage can occur if excessive infiltration/inflow (dilution by stormwater and/or groundwater) is allowed into a sanitary sewer system. Communities that have urbanized in the mid-20th century or later generally have built separate systems for sewage (sanitary sewers) and stormwater, because precipitation causes widely varying flows, reducing sewage treatment plant efficiency.

As rainfall travels over roofs and the ground, it may pick up various contaminants including soil particles and other sediment, heavy metals, organic compounds, animal waste, and oil and grease. Some jurisdictions require stormwater to receive some level of treatment before being discharged directly into waterways. Examples of treatment processes used for stormwater include retention basins, wetlands, buried vaults with various kinds of media filters, and vortex separators (to remove coarse solids).

Industrial Effluent

In highly regulated developed countries, industrial effluent usually receives at least pretreatment if not full treatment at the factories themselves to reduce the pollutant load, before discharge to the sewer. This process is called industrial wastewater treatment. The same does not apply to many developing countries where industrial effluent is more likely to enter the sewer if it exists, or even the receiving water body, without pretreatment.

Industrial wastewater may contain pollutants which cannot be removed by conventional sewage treatment. Also, variable flow of industrial waste associated with production cycles may upset the population dynamics of biological treatment units, such as the activated sludge process.

Process Steps

Overview

Sewage collection and treatment is typically subject to local, state and federal regulations and standards.

Treating wastewater has the aim to produce an effluent that will do as little harm as possible when discharged to the surrounding environment, thereby preventing pollution compared to releasing untreated wastewater into the environment.

Sewage treatment generally involves three stages, called primary, secondary and tertiary treatment.

- *Primary treatment* consists of temporarily holding the sewage in a quiescent basin where heavy solids can settle to the bottom while oil, grease and lighter solids float to the surface. The settled and floating materials are removed and the remaining liquid may be discharged or subjected to secondary treatment. Some sewage treatment plants that are connected to a combined sewer system have a bypass arrangement after the primary treatment unit. This means that during very heavy rainfall events, the secondary and tertiary treatment systems can be bypassed to protect them from hydraulic overloading, and the mixture of sewage and stormwater only receives primary treatment.

- *Secondary treatment* removes dissolved and suspended biological matter. Secondary treatment is typically performed by indigenous, water-borne micro-organisms in a managed habitat. Secondary treatment may require a separation process to remove the micro-organisms from the treated water prior to discharge or tertiary treatment.

- *Tertiary treatment* is sometimes defined as anything more than primary and secondary treatment in order to allow rejection into a highly sensitive or fragile ecosystem (estuaries, low-flow rivers, coral reefs,...). Treated water is sometimes disinfected chemically or physically (for example, by lagoons and microfiltration) prior to discharge into a stream, river, bay, lagoon or wetland, or it can be used for the irrigation of a golf course, green way or park. If it is sufficiently clean, it can also be used for groundwater recharge or agricultural purposes.

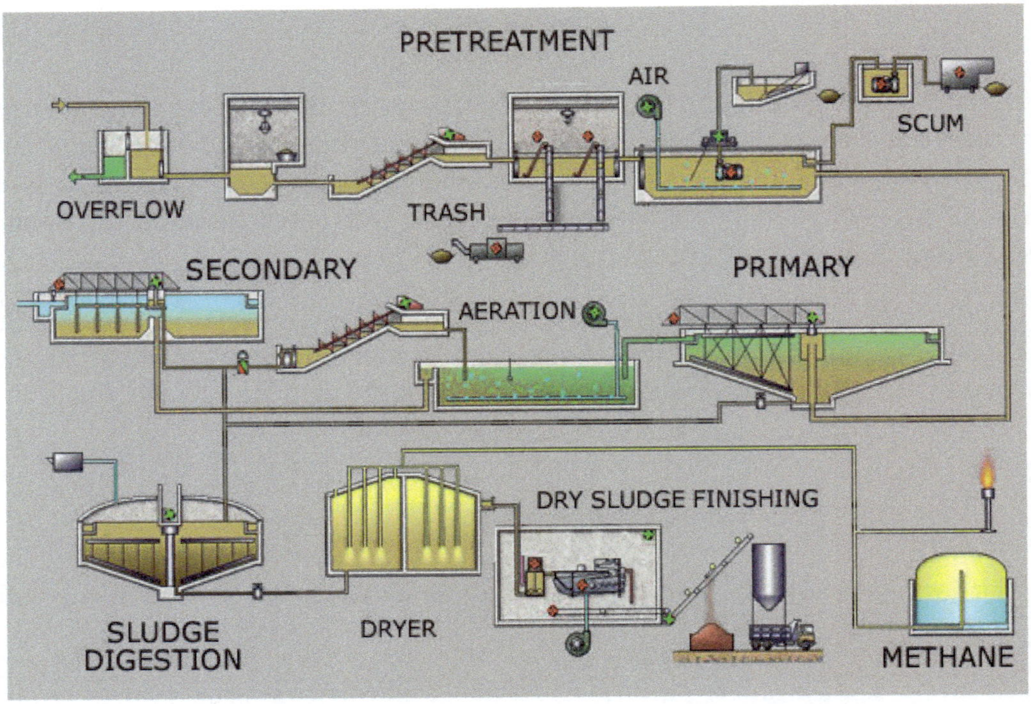

Simplified process flow diagram for a typical large-scale treatment plant

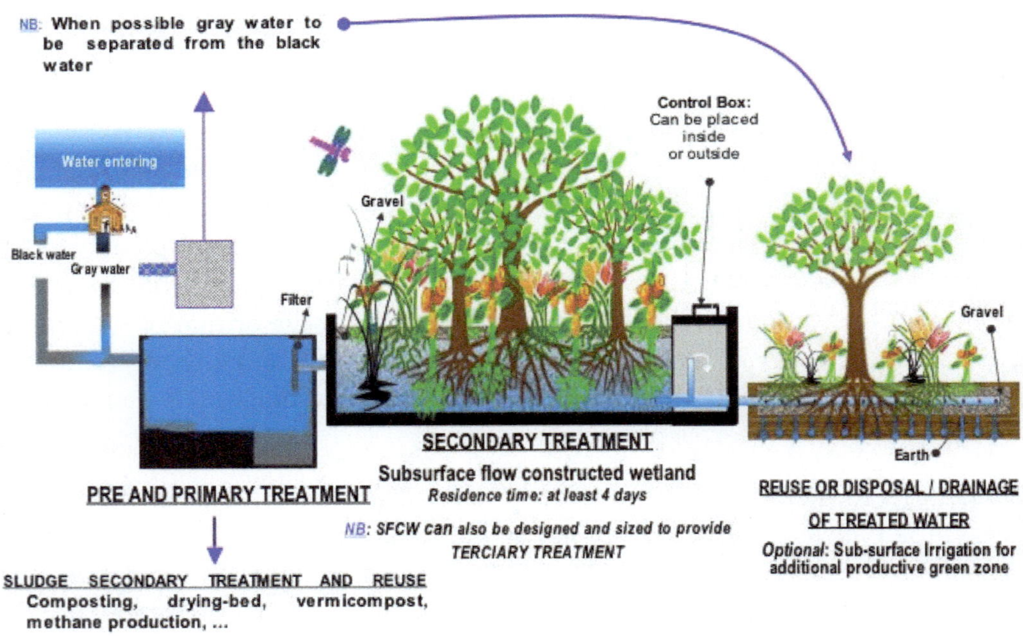

Process flow diagram for a typical treatment plant via subsurface flow constructed wetlands (SFCW)

Pretreatment

Pretreatment removes all materials that can be easily collected from the raw sewage before they damage or clog the pumps and sewage lines of primary treatment clarifiers. Objects commonly removed during pretreatment include trash, tree limbs, leaves, branches, and other large objects.

The influent in sewage water passes through a bar screen to remove all large objects like cans, rags, sticks, plastic packets etc. carried in the sewage stream. This is most commonly done with an automated mechanically raked bar screen in modern plants serving large populations, while in smaller or less modern plants, a manually cleaned screen may be used. The raking action of a mechanical bar screen is typically paced according to the accumulation on the bar screens and/or flow rate. The solids are collected and later disposed in a landfill, or incinerated. Bar screens or mesh screens of varying sizes may be used to optimize solids removal. If gross solids are not removed, they become entrained in pipes and moving parts of the treatment plant, and can cause substantial damage and inefficiency in the process.

Grit Removal

Pretreatment may include a sand or grit channel or chamber, where the velocity of the incoming sewage is adjusted to allow the settlement of sand, grit, stones, and broken glass. These particles are removed because they may damage pumps and other equipment. For small sanitary sewer systems, the grit chambers may not be necessary, but grit removal is desirable at larger plants. Grit chambers come in 3 types: horizontal grit chambers, aerated grit chambers and vortex grit chambers. The process is called sedimentation.

Flow Equalization

Clarifiers and mechanized secondary treatment are more efficient under uniform flow conditions. Equalization basins may be used for temporary storage of diurnal or wet-weather flow peaks. Basins provide a place to temporarily hold incoming sewage during plant maintenance and a means of diluting and distributing batch discharges of toxic or high-strength waste which might otherwise inhibit biological secondary treatment (including portable toilet waste, vehicle holding tanks, and septic tank pumpers). Flow equalization basins require variable discharge control, typically include provisions for bypass and cleaning, and may also include aerators. Cleaning may be easier if the basin is downstream of screening and grit removal.

Fat and Grease Removal

In some larger plants, fat and grease are removed by passing the sewage through a small tank where skimmers collect the fat floating on the surface. Air blowers in the base of the tank may also be used to help recover the fat as a froth. Many plants, however, use primary clarifiers with mechanical surface skimmers for fat and grease removal.

Primary Treatment

In the primary sedimentation stage, sewage flows through large tanks, commonly called “pre-settling basins”, “primary sedimentation tanks” or “primary clarifiers”. The tanks are used to settle sludge while grease and oils rise to the surface and are skimmed off. Primary settling tanks are usually equipped with mechanically driven scrapers that continually drive the collected sludge towards a hopper in the base of the tank where it is pumped to sludge treatment facilities.:[9–11] Grease and oil from the floating material can sometimes be recovered for saponification (soap making).

Primary treatment tanks in Oregon, USA.

Secondary Treatment

Secondary treatment is designed to substantially degrade the biological content of the sewage which are derived from human waste, food waste, soaps and detergent. The majority of municipal plants treat the settled sewage liquor using aerobic biological processes. To be effective, the biota require both oxygen and food to live. The bacteria and protozoa consume biodegradable soluble organic contaminants (e.g. sugars, fats, organic short-chain carbon molecules, etc.) and bind much of the less soluble fractions into floc. Secondary treatment systems are classified as *fixed-film* or *suspended-growth* systems.

- Fixed-film or attached growth systems include trickling filters, bio-towers, and rotating biological contactors, where the biomass grows on media and the sewage passes over its surface. The fixed-film principle has further developed into Moving Bed Biofilm Reactors (MBBR) and Integrated Fixed-Film Activated Sludge (IFAS) processes. An MBBR system typically requires a smaller footprint than suspended-growth systems.
- Suspended-growth systems include activated sludge, where the biomass is mixed with the sewage and can be operated in a smaller space than trickling filters that treat the same amount of water. However, fixed-film systems are more able to cope with drastic changes in the amount of biological material and can provide higher removal rates for organic material and suspended solids than suspended growth systems.

Secondary Sedimentation

Some secondary treatment methods include a secondary clarifier to settle out and separate biological floc or filter material grown in the secondary treatment bioreactor.

Secondary clarifier at a rural treatment plant.

List of Process Types

- Activated sludge
- Aerated lagoon
- Aerobic granulation
- Constructed wetland
- Membrane bioreactor
- Rotating biological contactor
- Sequencing batch reactor
- Trickling filter

To use less space, treat difficult waste, and intermittent flows, a number of designs of hybrid treatment plants have been produced. Such plants often combine at least two stages of the three main treatment stages into one combined stage. In the UK, where a large number of wastewater treatment plants serve small populations, package plants are a viable alternative to building a large structure for each process stage. In the US, package plants are typically used in rural areas, highway rest stops and trailer parks.

Tertiary Treatment

The purpose of tertiary treatment is to provide a final treatment stage to further improve the effluent quality before it is discharged to the receiving environment (sea, river, lake, wet lands, ground, etc.). More than one tertiary treatment process may be used at any treatment plant. If disinfection is practised, it is always the final process. It is also called "effluent polishing."

Filtration

Sand filtration removes much of the residual suspended matter. Filtration over activated carbon, also called *carbon adsorption,* removes residual toxins.

Lagoons or Ponds

A sewage treatment plant and lagoon in Everett, Washington, United States.

Lagoons or ponds provide settlement and further biological improvement through storage in large man-made ponds or lagoons. These lagoons are highly aerobic and colonization by native macrophytes, especially reeds, is often encouraged. Small filter-feeding invertebrates such as *Daphnia* and species of *Rotifera* greatly assist in treatment by removing fine particulates.

Biological Nutrient Removal

Biological nutrient removal (BNR) is regarded by some as a type of secondary treatment process, and by others as a tertiary (or "advanced") treatment process.

Wastewater may contain high levels of the nutrients nitrogen and phosphorus. Excessive release to the environment can lead to a buildup of nutrients, called eutrophication, which can in turn encourage the overgrowth of weeds, algae, and cyanobacteria (blue-green algae). This may cause an algal bloom, a rapid growth in the population of algae. The algae numbers are unsustainable and eventually most of them die. The decomposition of the algae by bacteria uses up so much of the oxygen in the water that most or all of the animals die, which creates more organic matter for the bacteria to decompose. In addition to causing deoxygenation, some algal species produce toxins that contaminate drinking water supplies. Different treatment processes are required to remove nitrogen and phosphorus.

Nitrogen Removal

Nitrogen is removed through the biological oxidation of nitrogen from ammonia to nitrate (nitrification), followed by denitrification, the reduction of nitrate to nitrogen gas. Nitrogen gas is released to the atmosphere and thus removed from the water.

Nitrification itself is a two-step aerobic process, each step facilitated by a different type of bacteria. The oxidation of ammonia (NH_3) to nitrite (NO_2^-) is most often facilitated by *Nitrosomonas* spp. ("nitroso" referring to the formation of a nitroso functional group). Nitrite oxidation to nitrate (NO_3^-), though traditionally believed to be facilitated by *Nitrobacter* spp. (nitro referring the formation of a nitro functional group), is now known to be facilitated in the environment almost exclusively by *Nitrospira* spp.

Denitrification requires anoxic conditions to encourage the appropriate biological communities to form. It is facilitated by a wide diversity of bacteria. Sand filters, lagooning and reed beds can all be used to reduce nitrogen, but the activated sludge process (if designed well) can do the job the most easily. Since denitrification is the reduction of nitrate to dinitrogen (molecular nitrogen) gas, an electron donor is needed. This can be, depending on the waste water, organic matter (from feces), sulfide, or an added donor like methanol. The sludge in the anoxic tanks (denitrification tanks) must be mixed well (mixture of recirculated mixed liquor, return activated sludge [RAS], and raw influent) e.g. by using submersible mixers in order to achieve the desired denitrification.

Sometimes the conversion of toxic ammonia to nitrate alone is referred to as tertiary treatment.

Over time, different treatment configurations have evolved as denitrification has become more sophisticated. An initial scheme, the Ludzack-Ettinger Process, placed an anoxic treatment zone before the aeration tank and clarifier, using the return activated sludge (RAS) from the clarifier as a nitrate source. Influent wastewater (either raw or as effluent from primary clarification) serves as the electron source for the facultative bacteria to metabolize carbon, using the inorganic nitrate as a source of oxygen instead of dissolved molecular oxygen. This denitrification scheme was naturally limited to the amount of soluble nitrate present in the RAS. Nitrate reduction was limited because RAS rate is limited by the performance of the clarifier.

The "Modified Ludzak-Ettinger Process" (MLE) is an improvement on the original concept, for it recycles mixed liquor from the discharge end of the aeration tank to the head of the anoxic tank to provide a consistent source of soluble nitrate for the facultative bacteria. In this instance, raw wastewater continues to provide the electron source, and sub-surface mixing maintains the bacteria in contact with both electron source and soluble nitrate in the absence of dissolved oxygen.

Many sewage treatment plants use centrifugal pumps to transfer the nitrified mixed liquor from the aeration zone to the anoxic zone for denitrification. These pumps are often referred to as *Internal Mixed Liquor Recycle* (IMLR) pumps. IMLR may be 200% to 400% the flow rate of influent wastewater (Q.) This is in addition to Return Activated Sludge (RAS) from secondary clarifiers, which may be 100% of Q. (Therefore, the hydraulic capacity of the tanks in such a system should handle at least 400% of annual average design flow (AADF.) At times, the raw or primary effluent wastewater must be carbon-supplemented by the addition of methanol, acetate, or simple food waste (molasses, whey, plant starch) to improve the treatment efficiency. These carbon additions should be accounted for in the design of a treatment facility's organic loading.

Further modifications to the MLE were to come: Bardenpho and Biodenipho processes include additional anoxic and oxidative processes to further polish the conversion of nitrate ion to molecular nitrogen gas. Use of an anaerobic tank following the initial anoxic process allows for luxury uptake of phosphorus by bacteria, thereby biologically reducing orthophosphate ion in the treated wastewater. Even newer improvements, such as Anammox Process, interrupt the formation of nitrate at the nitrite stage of nitrification, shunting nitrite-rich mixed liquor activated sludge to treatment where nitrite is then converted to molecular nitrogen gas, saving energy, alkalinity, and secondary carbon sourcing. Anammox™ (ANaerobic AMMonia OXidation) works by artificially extending detention time and preserving denitrifiying bacteria through the use of substrate added to the mixed liquor and continuously recycled from it prior to secondary clarification. Many other proprietary schemes are being deployed, including DEMON™, Sharon-ANAMMOX™, ANITA-Mox™, and DeAmmon™. The bacteria Brocadia anammoxidans can remove ammonium from waste water through anaerobic oxidation of ammonium to hydrazine, a form of rocket fuel.

Phosphorus Removal

Every adult human excretes between 200 and 1000 grams of phosphorus annually. Studies of United States sewage in the late 1960s estimated mean per capita contributions of 500 grams in urine and feces, 1000 grams in synthetic detergents, and lesser variable amounts used as corrosion and scale control chemicals in water supplies. Source control via alternative detergent formulations has subsequently reduced the largest contribution, but the content of urine and feces will remain unchanged. Phosphorus removal is important as it is a limiting nutrient for algae growth in many fresh water systems. It is also particularly important for water reuse systems where high phosphorus concentrations may lead to fouling of downstream equipment such as reverse osmosis.

Phosphorus can be removed biologically in a process called enhanced biological phosphorus removal. In this process, specific bacteria, called polyphosphate-accumulating organisms (PAOs), are selectively enriched and accumulate large quantities of phosphorus within their cells (up to 20 percent of their mass). When the biomass enriched in these bacteria is separated from the treated water, these biosolids have a high fertilizer value.

Phosphorus removal can also be achieved by chemical precipitation, usually with salts of iron (e.g. ferric chloride), aluminum (e.g. alum), or lime.[:18] This may lead to excessive sludge production as hydroxides precipitates and the added chemicals can be expensive. Chemical phosphorus removal requires significantly smaller equipment footprint than biological removal, is easier to operate and is often more reliable than biological phosphorus removal. Another method for phosphorus removal is to use granular laterite.

Once removed, phosphorus, in the form of a phosphate-rich sewage sludge, may be dumped in a landfill or used as fertilizer. In the latter case, the treated sewage sludge is also sometimes referred to as biosolids.

Disinfection

The purpose of disinfection in the treatment of waste water is to substantially reduce the number of microorganisms in the water to be discharged back into the environment for the later use of drinking, bathing, irrigation, etc. The effectiveness of disinfection depends on the quality of the water being treated (e.g., cloudiness, pH, etc.), the type of disinfection being used, the disinfectant dosage (concentration and time), and other environmental variables. Cloudy water will be treated less successfully, since solid matter can shield organisms, especially from ultraviolet light or if contact times are low. Generally, short contact times, low doses and high flows all militate against effective disinfection. Common methods of disinfection include ozone, chlorine, ultraviolet light, or sodium hypochlorite.[:16] Chloramine, which is used for drinking water, is not used in the treatment of waste water because of its persistence. After multiple steps of disinfection, the treated water is ready to be released back into the water cycle by means of the nearest body of water or agriculture. Afterwards, the water can be transferred to reserves for everyday human uses.

Chlorination remains the most common form of waste water disinfection in North America due to its low cost and long-term history of effectiveness. One disadvantage is that chlorination of residual organic material can generate chlorinated-organic compounds that may be carcinogenic or harmful to the environment. Residual chlorine or chloramines may also be capable of chlorinating organic material in the natural aquatic environment. Further, because residual chlorine is toxic to aquatic species, the treated effluent must also be chemically dechlorinated, adding to the complexity and cost of treatment.

Ultraviolet (UV) light can be used instead of chlorine, iodine, or other chemicals. Because no chemicals are used, the treated water has no adverse effect on organisms that later consume it, as may be the case with other methods. UV radiation causes damage to the genetic structure of bacteria, viruses, and other pathogens, making them incapable of reproduction. The key disadvantages of UV disinfection are the need for frequent lamp maintenance and replacement and the need for a highly treated effluent to ensure that the target microorganisms are not shielded from the UV radiation (i.e., any solids present in the treated effluent may protect microorganisms from the UV light). In the United Kingdom, UV light is becoming the most common means of disinfection because of the concerns about the impacts of chlorine in chlorinating residual organics in the wastewater and in chlorinating organics in the receiving water. Some sewage treatment systems in Canada and the US also use UV light for their effluent water disinfection.

Ozone (O_3) is generated by passing oxygen (O_2) through a high voltage potential resulting in a third oxygen atom becoming attached and forming O_3. Ozone is very unstable and reactive and oxidizes most organic material it comes in contact with, thereby destroying many pathogenic microorganisms. Ozone is considered to be safer than chlorine because, unlike chlorine which has to be stored on site (highly poisonous in the event of an accidental release), ozone is generated on-site as needed. Ozonation also produces fewer disinfection by-products than chlorination. A disadvantage of ozone disinfection is the high cost of the ozone generation equipment and the requirements for special operators.

Fourth Treatment Stage

Micropollutants such as pharmaceuticals, ingredients of household chemicals, chemicals used in small businesses or industries, environmental persistent pharmaceutical pollutant (EPPP) or pesticides may not be eliminated in the conventional treatment process (primary, secondary and tertiary treatment) and therefore lead to water pollution. Although concentrations of those substances and their decompostion products are quite low, there is still a chance to harm aquatic organisms. For pharmaceuticals, the following substances have been identified as "toxicologically relevant": substances with endocrine disrupting effects, genotoxic substances and substances that enhance the development of bacterial resistances. They mainly belong to the group of environmental persistent pharmaceutical pollutants. Techniques for elimination of micropollutants via a fourth treatment stage during sewage treatment are being tested in Germany, Switzerland and the Netherlands. However, since those techniques are still costly, they are not yet applied on a regular basis. Such process steps mainly consist of activated carbon filters that adsorb the micropollutants. Ozone can also be applied as an oxidative method. Also the use of enzymes such as the enzyme laccase is under investigation. A new concept which could provide an energy-efficient treatment of micropollutants could be the use of laccase secreting fungi cultivated at a wastewater treatment plant to degrade micropollutants and at the same time to provide enzymes at a cathode of a microbial biofuel cells. Microbial biofuel cells are investigated for their property to treat organic matter in wastewater.

To reduce pharmaceuticals in water bodies, also "source control" measures are under investigation, such as innovations in drug development or more responsible handling of drugs.

Odor Control

Odors emitted by sewage treatment are typically an indication of an anaerobic or "septic" condition. Early stages of processing will tend to produce foul-smelling gases, with hydrogen sulfide being most common in generating complaints. Large process plants in urban areas will often treat the odors with carbon reactors, a contact media with bio-slimes, small doses of chlorine, or circulating fluids to biologically capture and metabolize the noxious gases. Other methods of odor control exist, including addition of iron salts, hydrogen peroxide, calcium nitrate, etc. to manage hydrogen sulfide levels.

High-density solids pumps are suitable for reducing odors by conveying sludge through hermetic closed pipework.

Energy Requirements

For conventional sewage treatment plants, around 30 percent of the annual operating costs is usually required for energy. The energy requirements vary with type of treatment process as well as wastewater load. For example, constructed wetlands have a lower energy requirement than activated sludge plants, as less energy is required for the aeration step. Sewage treatment plants that produce biogas in their sewage sludge treatment process with anaerobic digestion can produce enough energy to meet most of the energy needs of the sewage treatment plant itself.

In conventional secondary treatment processes, most of the electricity is used for aeration, pumping systems and equipment for the dewatering and drying of sewage sludge. Advanced wastewater treatment plants, e.g. for nutrient removal, require more energy than plants that only achieve primary or secondary treatment.

Sludge Treatment and Disposal

The sludges accumulated in a wastewater treatment process must be treated and disposed of in a safe and effective manner. The purpose of digestion is to reduce the amount of organic matter and the number of disease-causing microorganisms present in the solids. The most common treatment options include anaerobic digestion, aerobic digestion, and composting. Incineration is also used, albeit to a much lesser degree.

Sludge treatment depends on the amount of solids generated and other site-specific conditions. Composting is most often applied to small-scale plants with aerobic digestion for mid-sized operations, and anaerobic digestion for the larger-scale operations.

The sludge is sometimes passed through a so-called pre-thickener which de-waters the sludge. Types of pre-thickeners include centrifugal sludge thickeners rotary drum sludge thickeners and belt filter presses. Dewatered sludge may be incinerated or transported offsite for disposal in a landfill or use as an agricultural soil amendment.

Environment Aspects

The outlet of the Karlsruhe sewage treatment plant flows into the Alb.

Many processes in a wastewater treatment plant are designed to mimic the natural treatment processes that occur in the environment, whether that environment is a natural water body or the ground. If not overloaded, bacteria in the environment will consume organic contaminants, although this will reduce the levels of oxygen in the water and may significantly change the overall ecology of the receiving water. Native bacterial populations feed on the organic contaminants, and the numbers of disease-causing microorganisms are reduced by natural environmental conditions such as predation or exposure to ultraviolet radiation. Consequently, in cases where the receiving environment provides a high level of dilution, a high degree of wastewater treatment may not be required. However, recent evidence has demonstrated that very low levels of specific contaminants in wastewater, including hormones (from animal husbandry and residue from human hormonal contraception methods) and synthetic materials such as phthalates that mimic hormones in their action, can have an unpredictable adverse impact on the natural biota and potentially on humans if the water is re-used for drinking water. In the US and EU, uncontrolled discharges of wastewater to the environment are not permitted under law, and strict water quality requirements are to be met, as clean drinking water is essential. A significant threat in the coming decades will be the increasing uncontrolled discharges of wastewater within rapidly developing countries.

Effects on Biology

Sewage treatment plants can have multiple effects on nutrient levels in the water that the treated sewage flows into. These nutrients can have large effects on the biological life in the water in contact with the effluent. Stabilization ponds (or sewage treatment ponds) can include any of the following:

- Oxidation ponds, which are aerobic bodies of water usually 1–2 meters in depth that receive effluent from sedimentation tanks or other forms of primary treatment.
    - ◇ Dominated by algae

- Polishing ponds are similar to oxidation ponds but receive effluent from an oxidation pond or from a plant with an extended mechanical treatment.
    - ◇ Dominated by zooplankton
- Facultative lagoons, raw sewage lagoons, or sewage lagoons are ponds where sewage is added with no primary treatment other than coarse screening. These ponds provide effective treatment when the surface remains aerobic; although anaerobic conditions may develop near the layer of settled sludge on the bottom of the pond.
- Anaerobic lagoons are heavily loaded ponds.
    - ◇ Dominated by bacteria
- Sludge lagoons are aerobic ponds, usually 2 to 5 meters in depth, that receive anaerobically digested primary sludge, or activated secondary sludge under water.
    - ◇ Upper layers are dominated by algae

Phosphorus limitation is a possible result from sewage treatment and results in flagellate-dominated plankton, particularly in summer and fall.

A phytoplankton study found high nutrient concentrations linked to sewage effluents. High nutrient concentration leads to high chlorophyll a concentrations, which is a proxy for primary production in marine environments. High primary production means high phytoplankton populations and most likely high zooplankton populations, because zooplankton feed on phytoplankton. However, effluent released into marine systems also leads to greater population instability.

The planktonic trends of high populations close to input of treated sewage is contrasted by the bacterial trend. In a study of *Aeromonas* spp. in increasing distance from a wastewater source, greater change in seasonal cycles was found the furthest from the effluent. This trend is so strong that the furthest location studied actually had an inversion of the *Aeromonas* spp. cycle in comparison to that of fecal coliforms. Since there is a main pattern in the cycles that occurred simultaneously at all stations it indicates seasonal factors (temperature, solar radiation, phytoplankton) control of the bacterial population. The effluent dominant species changes from *Aeromonas caviae* in winter to *Aeromonas sobria* in the spring and fall while the inflow dominant species is *Aeromonas caviae*, which is constant throughout the seasons.

Treated Sewage Reuse

With suitable technology, it is possible to reuse sewage effluent for drinking water, although this is usually only done in places with limited water supplies, such as Windhoek and Singapore.

In Israel, about 50 percent of agricultural water use (total use was 1 billion cubic metres in 2008) is provided through reclaimed sewer water. Future plans call for increased use of treated sewer water as well as more desalination plants.

Sewage Treatment in Developing Countries

Few reliable figures exist on the share of the wastewater collected in sewers that is being treated in the world. A global estimate by UNDP and UN-Habitat is that 90% of all wastewater generated is released into the environment untreated. In many developing countries the bulk of domestic and industrial wastewater is discharged without any treatment or after primary treatment only.

In Latin America about 15 percent of collected wastewater passes through treatment plants (with varying levels of actual treatment). In Venezuela, a below average country in South America with respect to wastewater treatment, 97 percent of the country's sewage is discharged raw into the environment. In Iran, a relatively developed Middle Eastern country, the majority of Tehran's population has totally untreated sewage injected to the city's groundwater. However, the construction of major parts of the sewage system, collection and treatment, in Tehran is almost complete, and under development, due to be fully completed by the end of 2012. In Isfahan, Iran's third largest city, sewage treatment was started more than 100 years ago.

Only few cities in sub-Saharan Africa have sewer-based sanitation systems, let alone wastewater treatment plants, an exception being South Africa and – until the late 1990s- Zimbabwe. Instead, most urban residents in sub-Saharan Africa rely on on-site sanitation systems without sewers, such as septic tanks and pit latrines, and faecal sludge management in these cities is an enormous challenge.

History

The Great Stink of 1858 stimulated research into the problem of sewage treatment. In this caricature in *The Times*, Michael Faraday reports to *Father Thames* on the state of the river.

Basic sewer systems were used for waste removal in ancient Mesopotamia, where vertical shafts carried the waste away into cesspools. Similar systems existed in the Indus Valley civilization in modern-day India and in Ancient Crete and Greece. In the Middle Ages the sewer systems built by the Romans fell into disuse and waste was collected into cesspools that were periodically emptied by workers known as 'rakers' who would often sell it as fertilizer to farmers outside the city.

Modern sewage systems were first built in the mid-nineteenth century as a reaction to the exacerbation of sanitary conditions brought on by heavy industrialization and urbanization. Due to the contaminated water supply, cholera outbreaks occurred in 1832, 1849 and 1855 in London, killing tens of thousands of people. This, combined with the Great Stink of 1858, when the smell of untreated human waste in the River Thames became overpowering, and the report into sanitation reform of the Royal Commissioner Edwin Chadwick, led to the Metropolitan Commission of Sewers appointing Sir Joseph Bazalgette to construct a vast underground sewage system for the safe removal of waste. Contrary to Chadwick's recommendations, Bazalgette's system, and others later built in Continental Europe, did not pump the sewage onto farm land for use as fertilizer; it was simply piped to a natural waterway away from population centres, and pumped back into the environment.

Early Attempts

One of the first attempts at diverting sewage for use as a fertilizer in the farm was made by the cotton mill owner James Smith in the 1840s. He experimented with a piped distribution system initially proposed by James Vetch that collected sewage from his factory and pumped it into the outlying farms, and his success was enthusiastically followed by Edwin Chadwick and supported by organic chemist Justus von Liebig.

The idea was officially adopted by the Health of Towns Commission, and various schemes (known as sewage farms) were trialled by different municipalities over the next 50 years. At first, the heavier solids were channeled into ditches on the side of the farm and were covered over when full, but soon flat-bottomed tanks were employed as reservoirs for the sewage; the earliest patent was taken out by William Higgs in 1846 for "tanks or reservoirs in which the contents of sewers and drains from cities, towns and villages are to be collected and the solid animal or vegetable matters therein contained, solidified and dried..." Improvements to the design of the tanks included the introduction of the horizontal-flow tank in the 1850s and the radial-flow tank in 1905. These tanks had to be manually de-sludged periodically, until the introduction of automatic mechanical de-sludgers in the early 1900s.

The precursor to the modern septic tank was the cesspool in which the water was sealed off to prevent contamination and the solid waste was slowly liquified due to anaerobic action; it was invented by L.H Mouras in France in the 1860s. Donald Cameron, as City Surveyor for Exeter patented an improved version in 1895, which he called a 'septic tank'; septic having the meaning of 'bacterial'. These are still in worldwide use, especially in rural areas unconnected to large-scale sewage systems.

Chemical Treatment

It was not until the late 19th century that it became possible to treat the sewage by chemically breaking it down through the use of microorganisms and removing the pollutants. Land treatment was also steadily becoming less feasible, as cities grew and the volume of sewage produced could no longer be absorbed by the farmland on the outskirts.

Sir Edward Frankland conducted experiments at the Sewage Farm in Croydon, England, during the 1870s and was able to demonstrate that filtration of sewage through porous gravel produced a nitrified effluent (the ammonia was converted into nitrate) and that the filter remained unclogged over long periods of time. This established the then revolutionary possibility of biological treatment of sewage using a contact bed to oxidize the waste. This concept was taken up by the chief chemist for the London Metropolitan Board of Works, William Libdin, in 1887:

> ...in all probability the true way of purifying sewage...will be first to separate the sludge, and then turn into neutral effluent... retain it for a sufficient period, during which time it should be fully aerated, and finally discharge it into the stream in a purified condition. This is indeed what is aimed at and imperfectly accomplished on a sewage farm.

From 1885 to 1891 filters working on this principle were constructed throughout the UK and the idea was also taken up in the US at the Lawrence Experiment Station in Massachusetts, where Frankland's work was confirmed. In 1890 the LES developed a 'trickling filter' that gave a much more reliable performance.

Sir Edward Frankland, a distinguished chemist, who demonstrated the possibility of chemically treating sewage in the 1870s.

Contact beds were developed in Salford, Lancashire and by scientists working for the London City Council in the early 1890s. According to Christopher Hamlin, this was part of a conceptual revolution that replaced the philosophy that saw "sewage purification as the prevention of decomposition with one that tried to facilitate the biological process that destroy sewage naturally."

Contact beds were tanks containing the inert substance, such as stones or slate, that maximized the surface area available for the microbial growth to break down the sewage. The sewage was held in the tank until it was fully decomposed and it was then filtered out into the ground. This method quickly became widespread, especially in the UK, where it was used in Leicester, Sheffield, Manchester and Leeds. The bacterial bed was simultaneously developed by Joseph Corbett as Borough Engineer in Salford and experiments in 1905 showed that his method was superior in that greater volumes of sewage could be purified better for longer periods of time than could be achieved by the contact bed.

The Royal Commission on Sewage Disposal published its eighth report in 1912 that set what became the international standard for sewage discharge into rivers; the '20:30 standard', which allowed 20 mg Biochemical oxygen demand and 30 mg suspended solid per litre.

Sewage Characteristics

Characterization of wastes is essential for an effective and economical waste management programme. It helps in the choice of treatment methods deciding the extent of treatment, assessing the beneficial uses of wastes and utilizing the waste purification capacity of natural bodies of water in a planned and controlled manner. While analysis of wastewater in each particular case is advisable, data from the other cities may be utilized during initial stage of planning.

Domestic sewage comprises spent water from kitchen, bathroom, lavatory, *etc*. The factors which contribute to variations in characteristics of the domestic sewage are daily per capita use of water, quality of water supply and the type, condition and extent of sewerage system, and habits of the people. Municipal sewage, which contains both domestic and industrial wastewater, may differ from place to place depending upon the type of industries and industrial establishment. The important characteristics of sewage are discussed here.

Temperature

The observations of temperature of sewage are useful in indicating solubility of oxygen, which affects transfer capacity of aeration equipment in aerobic systems, and rate of biological activity. Extremely low temperature affects adversely on the efficiency of biological treatment systems and on efficiency of sedimentation. In general, under Indian conditions the temperature of the raw sewage is observed to be between 15 and 35 ^{O}C at various places in different seasons.

The pH

The hydrogen ion concentration expressed as pH, is a valuable parameter in the operation of biological units. The pH of the fresh sewage is slightly more than the water supplied to the community. However, decomposition of organic matter may lower the pH, while the presence of industrial wastewater may produce extreme fluctuations. Generally the pH of raw sewage is in the range 5.5 to 8.0.

Colour and Odour

Fresh domestic sewage has a slightly soapy and cloudy appearance depending upon its concentration. As time passes the sewage becomes stale, darkening in colour with a pronounced smell due to microbial activity.

Solids

Though sewage generally contains less than 0.5 percent solids, the rest being water, still the nuisance caused by the solids cannot be overlooked, as these solids are highly degradable and therefore need proper disposal. The sewage solids may be classified into dissolved solids, suspended solids and volatile suspended solids. Knowledge of the volatile or organic fraction of solid, which decomposes, becomes necessary, as this constitutes the load on biological treatment units or oxygen resources of a stream when sewage is disposed off by dilution. The estimation of suspended solids, both organic and inorganic, gives a general picture of the load on sedimentation and grit removal system during sewage treatment. Dissolved inorganic fraction is to be considered when sewage is used for land irrigation or any other reuse is planned.

Nitrogen and Phosphorus

The principal nitrogen compounds in domestic sewage are proteins, amines, amino acids, and urea. Ammonia nitrogen in sewage results from the bacterial decomposition of these organic constituents. Nitrogen being an essential component of biological protoplasm, its concentration is important for proper functioning of biological treatment systems and disposal on land. Generally, the domestic sewage contains sufficient nitrogen, to take care of the needs of the biological treatment. For industrial wastewater if sufficient nitrogen is not present it is required to be added externally. Generally nitrogen content in the untreated sewage is observed to be in the range of 20 to 50 mg/L measured as TKN.

Phosphorus is contributing to domestic sewage from food residues containing phosphorus and their breakdown products. The use of increased quantities of synthetic detergents adds substantially to the phosphorus content of sewage. Phosphorus is also an essential nutrient for the biological processes. The concentration of phosphorus in domestic sewage is generally adequate to support aerobic biological wastewater treatment. However, it will be matter of concerned when the treated effluent is to be reused. The concentration of PO_4 in raw sewage is generally observed in the range of 5 to 10 mg/L.

Chlorides

Concentration of chlorides in sewage is greater than the normal chloride content of water supply. The chloride concentration in excess than the water supplied can be used as an index of the strength of the sewage. The daily contribution of chloride averages to about 8 gm per person. Based on an average sewage flow of 150 LPCD, this would result in the chloride content of sewage being 50 mg/L higher than that of the water supplied. Any abnormal increase should indicate discharge of chloride bearing wastes or saline groundwater infiltration, the latter adding to the sulphates as well, which may lead to excessive generation of hydrogen sulphide.

Organic Material

Organic compounds present in sewage are of particular interest for environmental engineering. A large variety of microorganisms (that may be present in the sewage or in the receiving water body) interact with the organic material by using it as an energy or material source. The utilization of the organic material by microorganisms is called metabolism. The conversion of organic material by microorganism to obtain energy is called catabolism and the incorporation of organic material in the cellular material is called anabolism.

To describe the metabolism of microorganisms and oxidation of organic material, it is necessary to characterize quantitatively concentration of organic matter in different forms. In view of the enormous variety of organic compounds in sewage it is totally unpractical to determine these individually. Thus a parameter must be used that characterizes a property that all these have in common. In practice two properties of almost all organic compounds can be used: (1) organic compound can be oxidized; and (2) organic compounds contain organic carbon.

In environmental engineering there are two standard tests based on the oxidation of organic material: 1) the Biochemical Oxygen Demand (BOD) and 2) the Chemical Oxygen Demand (COD) tests. In both tests, the organic material concentration is measured during the test. The essential differences between the COD and the BOD tests are in the oxidant utilized and the operational conditions imposed during the test such as biochemical oxidation and chemical oxidation. The other method for measuring organic material is the development of the Total Organic Carbon (TOC) test as an alternative to quantify the concentration of the organic material.

Biochemical Oxygen Demand (BOD): The BOD of the sewage is the amount of oxygen required for the biochemical decomposition of biodegradable organic matter under aerobic conditions. The oxygen consumed in the process is related to the amount of decomposable organic matter. The general range of BOD observed for raw sewage is 100 to 400 mg/L. Values in the lower range are being common under average Indian cities.

Chemical Oxygen Demand (COD): The COD gives the measure of the oxygen required for chemical oxidation. It does not differentiate between biological oxidisable and nonoxidisable material. However, the ratio of the COD to BOD does not change significantly for particular waste and hence this test could be used conveniently for interpreting performance efficiencies of the treatment units. In general, the COD of raw sewage at various places is reported to be in the range 200 to 700 mg/L.

In COD test, the oxidation of organic matter is essentially complete within two hours, whereas, biochemical oxidation of organic matter takes several weeks. In case of wastewaters with a large range of organic compounds, an extra difficulty in using BOD as a quantitative parameter is that the rate of oxidation of organic compounds depends on the nature and size of its molecules. Smaller molecules are readily available for use by bacteria, but large molecules and colloidal and suspended matters can only be metabolized after preparatory steps of hydrolysis. It is therefore not possible to establish a general relationship between the experimental five-day BOD and the ultimate BOD of a sample, *i.e.*, the oxygen consumption after several weeks. For sewage (with $k=0.23$ d^{-1} at $20^{0}C$) the BOD5 is 0.68 times of ultimate BOD, and ultimate BOD is 87% of the COD. Hence, the COD /BOD ratio for the sewage is around 1.7.

Toxic Metals and Compounds

Some heavy metals and compounds such as chromium, copper, cyanide, which are toxic may find their way into municipal sewage through industrial discharges. The concentration of these compounds is important if the sewage is to treat by biological treatment methods or disposed off in stream or on land. In general these compounds are within toxic limits in sanitary sewage; however, with receipt of industrial discharges they may cross the limits in municipal wastewaters.

Effect of Industrial Wastes

Wastewaters from industries can form important component of sewage in both volume and composition. It is therefore necessary that details about nature of industries, the quantity and characteristics of the wastewater and their variations, which may affect the sewerage system and sewage treatment process, should be collected.

In case, where wastewaters high in suspended solids and BOD are to be accepted, provision should be made in the design of the treatment plant to handle such wastes. In certain instances, it is more economical to tackle the industrial waste at the source itself. Where, the wastewater has high or low pH, corrective measures are necessary before admitting them to the sewers or the treatment plant. Toxic metals and chemicals having adverse effects on biological treatment processes, or upon fish life in a natural water course, or render the receiving water stream unfit as a source of water supply, should be brought down to acceptable limits at the source itself. Oil and grease in excessive amounts not only add considerably to the cost of treatment, but also pose a disposal problem. The industrial wastewaters may be discharged into public sewers if the effluents meet the tolerance limits prescribed by the authority. If the wastewaters are to be discharged into inland surface waters, tolerance limits set by the concerned authority should be satisfied.

Effluent Disposal and Utilization

The sewage after treatment may be disposed either into a water body such as lake, stream, river, estuary, and ocean or on to land. It may also be utilized for several purposes such as (a) industrial reuse or reclaimed sewage effluent cooling system, boiler feed, process water, *etc.*, (b) reuse in agriculture and horticulture, watering of lawns, golf courses and similar purpose, and (c) groundwater recharge for augmenting groundwater resources for downstream users or for preventing saline water intrusion in coastal areas.

Status of Wastewater Generation, Collection, and Treatment in Indian Metro Cities

The prime cause of critical unsanitary conditions in many cities in India is due the lack of facilities to collect wastewater and to dispose off after treatment. Data on wastewater generation and collection is less when compared to information on water supply. Hence, it is difficult to assess the total pollution potential. As per the CPCB reports the total wastewater generated by 23 metro cities is 9,275 MLD [CPCB, 1997]. Out of this, about 58.5% is generated by the first four metro cities, *viz.* Bombay, Calcutta, Delhi and Chennai. The city of Bombay generates the maximum wastewater to the tune of 2,456 MLD and Madurai generates the least with 48 MLD [CPCB, 1997]. From the available data it may be seen that the ratio of industrial to municipal wastewater varies from 0.06% to 2%. Out of the 23 metrocities, 19 cities have sewerage coverage for more than 75% of the population and the remaining 4 cities have more than 50% coverage. On the whole 78% of the total metro population is provided with sewerage facility, compared to 63% in 1988 [CPCB, 1997].

Out of 9275 MLD of total wastewater generated, only 31% (2,923 MLD) is treated before letting out and the rest *i.e.*, 6,352 MLD is disposed off untreated. Three cities have only primary treatment facilities and thirteen have primary and secondary facilities. The municipalities dispose off their treated or partly treated or untreated wastewater into natural drains joining rivers or lakes or used on land for irrigation or fodder cultivation or into the sea or combination thereof.

It is found that in 12 metrocities there is some level of organized sewage farming under the control of government or local body. The municipal corporations of Bhopal, Calcutta, Hyderabad, Indore, Jaipur, Madras, Nagpur, Patna, Pune, Surat, Vadodara and Varanasi have sewage farms organized by government / farmers and controlled by Government / Municipal Corporation / irrigation departments. The cost of sewage charge was in the range of Rs.400/ hectare / year in Jaipur to Rs.75/hectare / year in Hyderabad. The average sale price of sewage works out to be Rs.188/ hectare / year for metrocities.

Economic Value of Sewage

The sewage contains nutrients, which if not optimally reused may cause eutrophication in receiving water bodies, thus causing their premature ageing. Hence, instead of directly discharging the effluents into water bodies it can be used for irrigation or fodder cultivation. The economic value of sewage can be assessed based on its nutrient value. This will guide for considering sewage as a source of income, and to make sewage treatment economically viable.

The nutrient value of sewage in terms of nitrogen 30 mg/L, phosphate 7.5 mg/L, and potassium 25 mg/L is provided by CPCB . The total value of nutrient in sewage assuming @ Rs. 4220/- per tone of nutrient (as per 1996 cost), works out to be Rs. 1018 million, *i.e.*, Rs. 890.6 million towards nutrients plus Rs. 127.4 million toward the cost of water.

A realistic rate for tariff towards sewage supplied for sewage farming should consider the cost of nutrients apart from the cost of water supplied. At present the sewage is charged at average rate of Rs. 188/hectare/ annum, which is towards the cost of irrigation water only. If nutrients in the sewage are also to be accounted for, then an additional cost of Rs. 263/MLD or Rs. 1315 per hectare/annum should be levied for application levels of 500 cm per hectare per annum. Hence, the tariff should be levied at Rs. 1503 per hectare/annum (Rs.1315 + 188) from cultivators [CPCB, 1997].

Classification and Application of Wastewater Treatment Methods

The degree of treatment required can be determined by comparing the influent wastewater characteristics to the required effluent characteristics, adhering to the regulations. Number of different treatment alternatives can be developed to achieve the treated wastewater quality.

Classification of Treatment Methods

The individual treatment methods are usually classified as:

- Physical unit operations
- Chemical unit processes
- Biological unit processes.

Physical Unit Operations: Treatment methods in which the application of physical forces predominates are known as physical unit operations. Most of these methods are based on physical forces, e.g. screening, mixing, flocculation, sedimentation, flotation, and filtration.

Chemical Unit Processes: Treatment methods in which removal or conversion of contaminant is brought by addition of chemicals or by other chemical reaction are known as chemical unit processes, for example, precipitation, gas transfer, adsorption, and disinfection.

Biological Unit Processes: Treatment methods in which the removal of contaminants is brought about by biological activity are known as biological unit processes.

- This is primarily used to remove biodegradable organic substances from the wastewater, either in colloidal or dissolved form.
- In the biological unit process, organic matter is converted into gases that can escape to the atmosphere and into bacterial cells, which can be removed by settling.
- Biological treatment is also used for nitrogen removal and for phosphorous and sulphate removal from the wastewater.

The different treatment methods used in wastewater treatment plant are classified in three different categories as:

- *Primary Treatment* : Refers to physical unit operations.
- *Secondary Treatment*: Refers to chemical and biological unit processes.
- *Tertiary Treatment*: Referes to any one or combination of two or all three i.e., physical unit operations and chemical or biological unit processes, used after secondary treatment.

Elements of Plant Analysis and Design

The important terms used in analysis and design of treatment plants are (CPHEEO, 1993): *Flow Sheet*: It is the graphical representation of a particular combination of unit operations and processes used in treatment.

Process Loading Criteria (or designed criteria): The criteria used as the basis for sizing the individual unit operation or process is known as process loading criteria.

Solid Balance: It is determined by identifying the quantities of solids entering and leaving each unit operation or process.

Hydraulic profile: This is used to identify the elevation of free surface of wastewater as it flows through various treatment units.

Plant Layout: It is spatial arrangement of the physical facilities of the treatment plant identified in the flow sheet.

Order of Reaction

The reactions occurring during wastewater treatment are slow and hence, kinetic considerations are important for design. The general equation used for relating the rate of change of concentration with respect to time can be expressed as

$$dS/dt = K.\ S^n$$

Where, S is the concentration of the reacting substance, K is the reaction rate constant per unit time, and n denotes the order of the reaction (n = 1 for first order reaction, n = 2 for second order reaction, and so on).

The value of K depends on the environmental conditions in the reactor, such as (a) temperature, (b) presence of toxicity, (c) presence of catalyst, (d) availability of nutrients and growth factors.

Zero order reactions (n = 0) are independent of the substance concentration and hence their rate (dS/dt) is constant. Certain catalytic reactions occur in this way and some times even biological reaction may follow zero order reaction.

In first order reactions, the rate of change of concentration of substance is proportional to the concentration of that substance. This concentration of the substance and rate will diminish with respect to time. Decomposition of single substrate exhibits the true first order reaction.

Biological stabilization of organic matter in batch reactor is a typical example of a pseudo- first-order reaction. The rate of reaction is proportional to the concentration of a single item, organic matter in this case, provided the other parameters controlling reactions are favourable. If the substrate concentration (organic matter) is maintained constant within the narrow range (as in the case of continuous flowing, completely mixed reactors), then the rate of reaction is practically constant and then it is like pseudo-zero-order type of reaction. Some biological treatment systems behave in this manner.

There are various complex processes whose overall rate is approximately first order in nature. With a complex substrates (sewage or industrial wastewaters) over all reaction rate may appear like a first order reaction, although the individual substrate among the several may exhibit the zero order reaction. This is because, the rate of reaction may be higher initially due to higher utilization of easily biodegradable substrate, but rate will slower down with respect to time due to more difficult substrate left in the reactor.

Types of Reactors Used

a) *Batch Reactor*: These reactors are operated as fill and draw type. In this the wastewater flow is not continuous in the reactor. The reactors are operated in batch mode with fill time, reaction time, and withdrawal time. For example, BOD test, Sequencing Batch Reactor (SBR). The reactor content may be completely mixed to ensure that no temperature or concentration gradient exists. All the elements in the reactor, under batch mode of operation, are exposed to treatment for the same length of time for which the substrate is held in the reactor. Hence, they are like ideal plug flow reactors.

b) *Plug-Flow (tubular flow) Reactor*: In this reactor, the fluid particles pass through the tank and are discharged in the same sequence in which they enter in the tank. The particles remain in the tank for a time equal to theoretical detention time. There is no overtaking or falling behind;

no intermixing or dispersion. Longitudinal dispersion is considered as minimum and this type can occur in high length to width ratio of the tanks. For example, grit chamber, aeration tank of ASP with high length to width ratio.

c) *Continuous-flow Stirred Tank (Complete – mixed) reactor*: In this reactors, particles are dispersed immediately throughout the tank as they enter the tank. Thus, the content in the reactor are perfectly homogeneous at all points in the reactor. This can be achieved in square, circular or rectangular tank. The particles leave the tank in proportion to their statistical population. The concentration of the effluent from the reactor is the same as that in the reactor.

d) *Arbitrary Flow*: Any degree of partial mixing between plug flow and completely mixing condition exists in this reactor. Each element of the incoming flow resides in the reactor for different length of time. It is also called as intermixing or dispersed flow and lies between ideal plug flow and ideal completely mixed reactor. This flow condition can be used in practice to describe the flow conditions in most of the reactors.

e) *Packed Bed Reactor*: They are filled with some packing medium, such as, rock, slag, ceramic or synthetic plastic media. With respect to flow they can be anaerobic filter, when completely filled and no air is supplied, or aerobic (trickling filter) when flow is intermittent or submerged aerobic filter when compressed air is supplied from the bottom.

f) *Fluidized Bed Reactor*: This reactor is similar to packed bed except packing medium is expanded by upward movement of fluid (or air) than resting on each other in fixed bed. The porosity or degree of fluidization can be controlled by controlling flow rate of fluid (wastewater or air).

Flow Patterns of Reactors

The flow pattern in the reactors depends on mixing conditions in them. This mixing in tern depends upon the shape of the reactor, energy spent per unit volume of the reactor, the size and scale of the unit, up-flow velocity of the liquid, rate of biogas generation (in an anaerobic reactors) or the rate of gas supplied (in an aerobic reactor), etc. Flow pattern affect the time of exposure to treatment and substrate distribution in the reactor. Depending upon the flow pattern the reactors can be classified as:

1. Batch reactors,
2. Ideal plug flow reactors,
3. Ideal completely-mixed flow reactors,
4. Non –ideal, dispersed flow reactors, and
5. Series or parallel combinations of the reactors.

The hydraulic regime in the reactor can be defined with respect to the 'Dispersion number', which characterizes mixing condition in the reactor (Arceivala and Asolekar, 2007).

Dispersion Number = D/UL

Where,

D = Axial or longitudinal dispersion coefficient, L^2/t

U = Mean flow velocity along the reactor, L/t

L = Length of axial travel path, L

For ideal plug flow D/UL = 0, since, dispersion is zero by definition.

D/UL ≤ 0.2 indicate the regime approaching plug flow conditions.

D/UL ≥ 3.0 to 4.0 indicates approaching completely mixed conditions.

Sewage Treatment Flow Sheet

The design of process flow sheet involves selection of an appropriate combination of various unit operations and unit processes to achieve a desired degree of contaminant removal. The selection of unit operations and processes primarily depends on the characteristics of the sewage and the required level of contaminants permitted in the treated effluents. The design of process flow sheet is important step in overall design of wastewater treatment and requires thorough understanding of the treatment units. It calls for optimization of wastewater treatment system coupled with stage wise optimal design of individual operation/ process to achieve a minimal cost design.

The main contaminants in domestic sewage, to be removed, are biodegradable organics, Suspended Solids (SS) and pathogens, with first two having been considered as the performance indicators for various treatment units. In general the objective of the domestic wastewater treatment is to bring down BOD less than 30 mg/L and SS less than 30 mg/L for disposal into inland water bodies.

The conventional flow sheet of sewage treatment plant consists of unit operations such as screening, grit removal, and Primary Settling Tank (PST), followed by unit process of aerobic biological treatment such as Activated Sludge Process (ASP) or Trickling Filter. The sludge removed from primary and secondary sedimentation tanks are digested anaerobically followed by drying of anaerobically digested sludge on sand drying beds. This process flow sheet is presented in Figure.

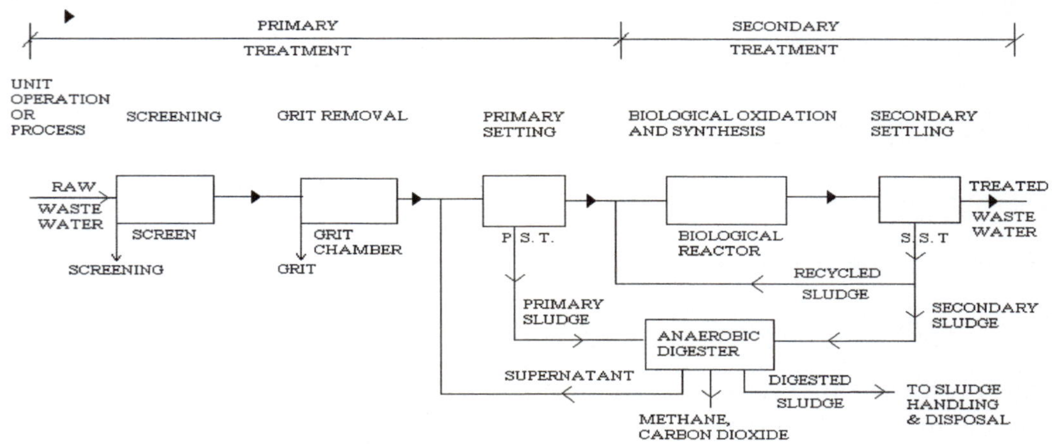

Process Flow-sheet of Conventional Domestic Sewage Treatment Plant

It is possible to replace the activated sludge process or trickling filter process by low cost treatment devices such as oxidation ditch, aerated lagoon or waste stabilization ponds. Such treatment devices obviate the necessity of some of the unit operations and processes like primary sedimentation and anaerobic digestion. Some of the process flow sheets are shown in Figure.

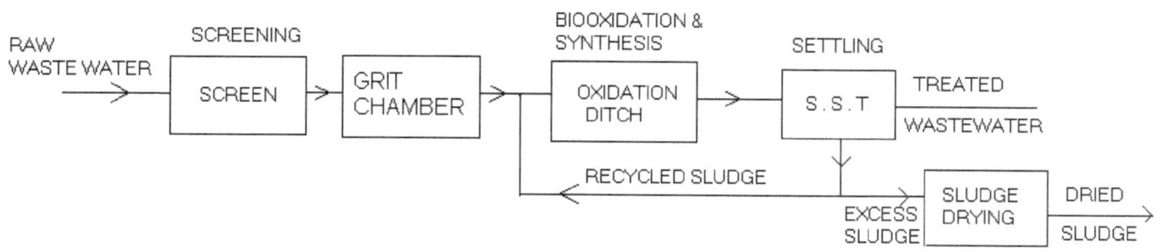

a) Process Flow sheet Incorporating Oxidation Ditch

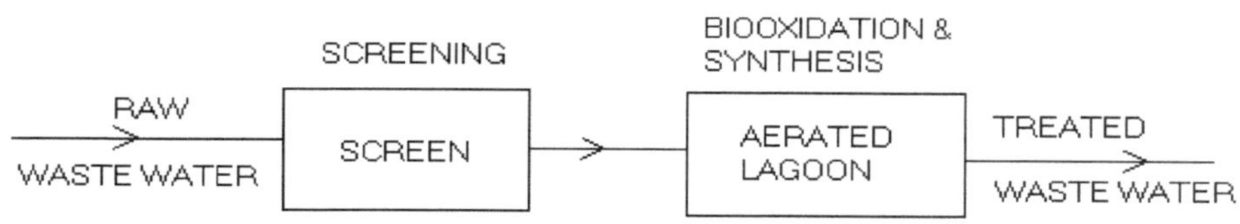

b) Process Flow sheet Employing Aerated Lagoon

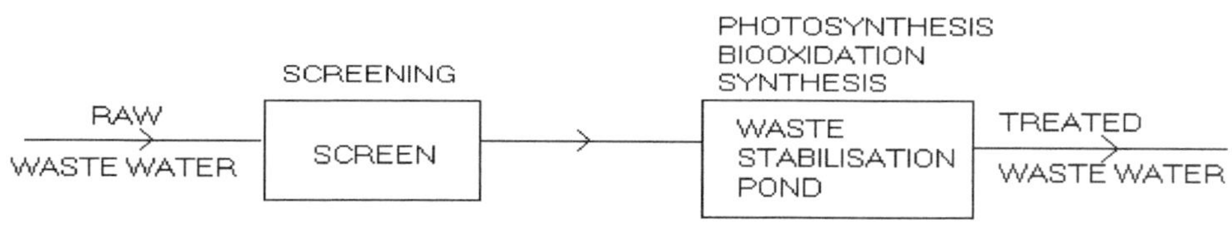

c) Process Flow sheet Employing Waste Stabilization Pond

Process flow sheet using oxidation ditch, aerated lagoon, and waste stabilization pond

With the better understanding of microbiology and biochemistry of anaerobic treatment, it is now feasible to treat dilute organic wastewater such as domestic wastewater directly through anaerobic treatment using recently developed innovative device such as Upflow Anaerobic Sludge Blanket Reactor (UASBR), Fluid–Bed Submerged Media Anaerobic Reactor (FB- SMAR) and Anaerobic Filter (AF) or Static–Bed SMAR (SB-SMAR) and Anaerobic Rotating Biological Contactor (An-RBC). Though, enough field data is to be generated as yet on their performance, it is generally reported that BOD5 removal efficiencies may range from 60-80%. Consequently, post treatment will generally be required to achieve the prescribed effluent standards. The process flow sheet anaerobic process is depicted in Figure.

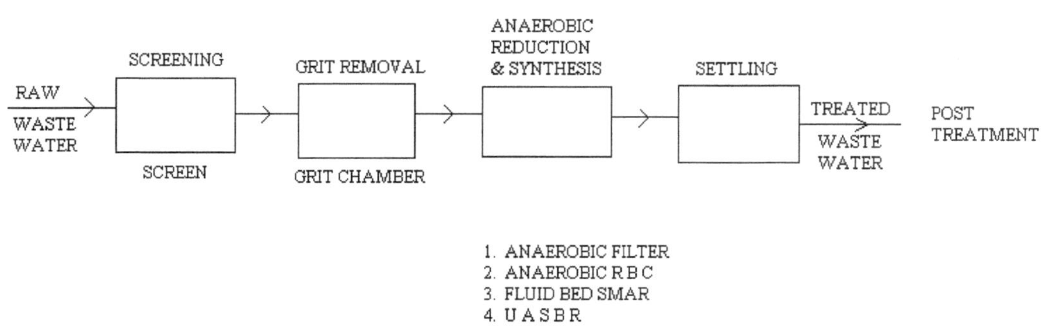

Process flow sheet employing anaerobic treatment system (CPHEEO, 1993)

Primary Treatment Units

Primary treatment consists solely separating the floating materials and also the heavy settable organic and inorganic solids. It also helps in removing the oils and grease from the sewage. This treatment reduces the BOD of the wastewater by about 15 to 30%. The operations used are screening for removing floating papers, rages, cloths, etc., grit chambers or detritus tanks for removing grit and sand, and skimming tanks for removing oils and grease; and primary settling tank is provided for removal of residual suspended matter. The organic solids, which are separated out in the sedimentation tanks in primary treatment, are often stabilized by anaerobic decomposition in digestion tank or incinerated. After digestion the sludge can be used as manure after drying on sludge drying beds or by some other means.

Bar Screens

Bar screen is a set of inclined parallel bars, fixed at a certain distance apart in a channel. These are used for removing larger particles of floating and suspended matter. The wastewater entering the screening channel should have a minimum self-clearing velocity 0.375 m/sec. Also the velocity should not rise to such extent as to dislodge the screenings from the bars. The slope of the hand-cleaned screens should be between 30° and 45° with the horizontal and that of mechanically cleaned screens may be between 45° and 80°. The submerged area of the surface of the screen, including bars and opening should be about 200% of the c/s area of the extract sewer for separate sewers and 300% for combined sewers. Clear spacing of bars for hand cleaned bar screens may be from 25 to 50 mm and that for mechanically cleaned bars may range from 15 mm to 75 mm. The width of the bars, facing the flow may be 8 mm to 15 mm and depth may vary from 25 mm to 75 mm, but sizes less than 8 x 25 mm are normally not used.

Grit Chamber

Grit chambers are designed to remove grit consists of sand, gravel, cinders or other inert solid materials that have specific gravity about 2.65, which is much greater than those of the organic solids in the wastewater. In this chamber particles settle as individual entities and there is no significant interaction with the neighboring particles. This type of settling is referred as free settling or zone-I settling. For proper functioning of the grit chamber, the velocity through the grit chamber should not be allowed to change in spite of the change in flow. One of the most satisfactory types of automatic velocity control is achieved by providing a proportional weir at the outlet.

The horizontal flow grit chambers should be designed in such a way that under the most adverse conditions, all the grit particles of size 0.20 mm or more in diameter should reach the bed of the channel prior to reaching outlet end. The length of the channel depends on the depth required which again depends on the settling velocity. A minimum allowance of approximately twice the maximum depth should be given for inlet and outlet zones. An allowance of 20-50% of the theoretical length of the channel may also be given.

Width of grit chamber should be between 1 m to 1.5 m and depth of flow is normally kept shallow. For total depth of channel a free board of about 0.3 m and grit space about 0.25 m should be provided. For larger plants two or more number of grit chambers in parallel may be used. In grit chambers the recommended detention time is about 30 to 60 seconds.

Skimming Tank

The floating solid materials such as soap, vegetables, debris, fruit skins, pieces of corks, *etc.* and oil and grease are removed from the wastewater in skimming tanks. A skimming tank is a chamber designed so that floating matter rises and remains on surface of the wastewater until removed, while the liquid flows continuously through outlet or partition below the water lines. The detention time in skimming tank is 3 minutes. To prevent heavy solids from settling at the bed, compressed air is blown through the diffusers placed in the floor of the tank. Due to compress air supply, the oily matters rise upward and are collected in the side trough, from where they are removed. In conventional sewage treatment plant separate skimming tank is not used and these materials are removed by providing baffle ahead of the effluent end of the primary sedimentation tank.

Primary Sedimentation Tank

Effluent of the grit chamber, containing mainly lightweight organic matter, is settled in the primary sedimentation tanks. The objective of treatment by sedimentation is to remove readily settleable solids and floating material and thus to reduce the suspended solids content when they are used as preliminary step to biological treatment, their function is to reduce the load on the biological treatment units.

The primary sedimentation tanks are usually designed for a flow through velocity of 1 cm/sec at average rate of flow. The detention period in the range of 90 to 150 minutes may be used for design. These tanks may be square, circular, or rectangular in plan with depth varying from 2.3 to 5 m. The diameter of circular tanks may be up to 40 m. The width of rectangular tank may be 10 to 25 m and the length may be up to 100 m. But to avoid water currents due to wind, length is limited up to 40 m. The slope of sludge hoppers in these tanks is generally 2:1 (vertical: horizontal). The slope of 1% is normally provided at the bed for rectangular tanks and 7.5 to 10% for circular tanks. This slope is necessary so that solids may slide to the bottom by gravity.

Secondary Treatment

The effluent from primary treatment is treated further for removal of dissolved and colloidal organic matter in secondary treatment. This is generally accomplished through biochemical decomposition of organic matter, which can be carried out either under aerobic or anaerobic conditions. In these biological units, bacteria's decompose the fine organic matter, to produce clearer effluent. The end products of aerobic decomposition are mainly carbon dioxide and bacterial cells, and that for anaerobic process are CH_4, CO_2 and bacterial cells.

The biological reactor in which the organic matter is decomposed (oxidized) by aerobic bacteria may consist of:

1) Filters (tricking filters),

2) Activated Sludge Process (ASP),

3) Oxidation ponds, *etc.*

The bacterial cells separated out in secondary setting tanks will be disposed after stabilizing them under aerobic or anaerobic process in a sludge digestion tank along with the solids settled in primary sedimentation tanks.

Trickling Filter

Trickling filters can be used for complete treatment for domestic waste and as roughing filter for strong industrial waste prior to activated sludge process. The primary sedimentation tank is provided prior to trickling filter so that the settleable solids in the sewage may not clog the filter. The trickling filter is followed by secondary settling tank for removal of settleable biosolids produced in filtration process.

As the wastewater trickles through the filter media (consisting rocks of 40 to 100 mm size or plastic media), a biological slime consisting of aerobic bacteria and other biota builds up around the media surface. Organic material in the sewage is absorbed on the biological slime, where they are partly degraded by the biota, thus increasing the thickness of the biofilm. Eventually there is a scouring of the biofilm and fresh biofilm begins to grow on the media. This phenomenon of detachment of the biofilm is called sloughing of the filter.

The trickling filters are classified as low rate and high rate depending on the organic and hydraulic loadings. Low rate filters are designed for hydraulic loading of 1 to 4 m^3/m^2.d and organic loadings as 80 to 320 g BOD/m^3.d. The high rate trickling filters are designed for hydraulic loading of 10 to 30 m^3/m^2.d (including recirculation) and organic loading of 500 to 1000 g BOD/m^3.d (excluding recirculation). Generally recirculation is not adopted in low rate filter and recirculation ratio of 0.5 to 3.0 or higher is used in case of high rate trickling filters. The depth of media varies from 1.0 to 1.8 m for high rate filters and 2.0 to 3.0 m for low rate filters. The bed of trickling filter is provided with slope 1 in 100 to 1 in 50. The under drainage system consists of 'V' shaped or half round channels, cast in concrete floor during its construction. Revolving distributors are provided at top with two or four horizontal arms of the pipe having perforations or holes. These rotating arms remain 15 to 25 cm above the top surface of the media. The distribution arms are rotated by the electric motor or by back reaction on the arms by the wastewater, at about 2 rpm. The head of 30 to 80 cm of wastewater is required to rotate the arms.

Activated Sludge Process

It is aerobic biological treatment system. The settled wastewater is aerated in an aeration tank for a period of few hours. During the aeration, the microorganisms in the aeration tank stabilize the organic matter. In this process part of the organic matter is synthesized into new cells and part is oxidized to derive energy. The synthesis reaction followed by subsequent separation of the resulting biological mass and the oxidation reaction is the main mechanism of BOD removal in the activated sludge process.

The biomass generated in the aeration tank is generally flocculent and it is separated from the aerated wastewater in a secondary settling tank and is recycled partially to the aeration tank. The mixture of recycled sludge and wastewater in the aeration tank is referred as mixed liquor. The recycling of sludge helps in the initial built up of a high concentration of active microorganism in the mixed liquor, which accelerates BOD removal. Once the required concentration of microorganism in the mixed liquor has been reached its further increase is prevented by the regulating quantity of sludge recycled and wasting the excess sludge from the system.

Aeration units are main units of activated sludge process, the main aim of which is to supply oxygen to the wastewater to keep the reactor content aerobic and to mix up the return sludge with wastewater thoroughly. The usual practice is to keep the detention period between 6 to 8 hours for treatment of sewage or similar industrial wastewater. The volume of aeration tank is also decided by considering the return sludge, which is about 25 to 50% of the wastewater volume.

Normally liquid depth provided should be between 3 and 4.5 m. A free board of 0.3 to 0.6 m is also provided. The mode of air supply in aeration tank can be either diffused air aeration, by supplementing compressed air from tank bottom, or by mechanical aerators provided at surface or by both diffused aeration and mechanical aerators. Depending on flow regime the activated sludge process can be classified as conventional (plug flow) and completely mixed activated sludge process. The modification of activated sludge process such as extended aeration is popularly used for treatment of wastewaters. The extended aeration is design for higher hydraulic retention time (18 h) and low F/M ratio.

Secondary Settling Tank (SST)

Design of secondary settling is somewhat different than that of the primary settling tanks. In the secondary settling tank the function served is clarification as well as thickening of the sludge. This type of settling which takes place in secondary settling tank is refereed as zone settling followed by compression. The SST is designed for detention period of 1.5 to 2.5 h. The depth of the tank can be between 2.5 and 4.5 m. The area of the tank is to worked out on the basis of surface overflow rate, overflow rate for SST of trickling filter should be 15-25 m^3/m^2.d and for SST of ASP 15-35 m^3/m^2.d at average flow. The length of effluent weir should be such that the weir loading rate is less than 185 m^3/m.d.

Oxidation Ponds

Oxidation ponds are the stabilization ponds, which received partially treated sewage. It is an earthen pond dug into the ground with shallow depth. The pond should be at least 1.0 m deep to discourage growth of aquatic weeds and should not exceed 1.8 m. The detention time in the pond is usually 1 to 4 weeks depending upon sunlight and temperature. Better efficiency of treatment is obtained if several ponds are placed in series so that the sewage flows progressively from one to another unit until it is finally discharged.

The surface area of the pond may be worked out by assuming a suitable value of organic loading which may range from 150 –300 kg/ha/d in hot tropical countries like India. Each unit may have an area ranging between 0.5 to 1.0 hectare.

The length of the tank may be kept about twice the width. A free board of about 1 m may also be provided above a capacity corresponding to 20-30 days of detention period. Properly operated ponds may be as effective as trickling filter in reducing the BOD of sewage. The BOD removal efficiency of pond is up to 90% and Coliform removal efficiency of pond is up to 99%.

Sludge Treatment

Sludge drying beds are commonly used in small wastewater treatment plants to dewater the sludge prior to final disposal. Two mechanisms are involved in the process, such as filtration of water through the sand, and evaporation of water from sludge surface. The filtered water is returned to the plant for treatment. The process is well suited to sludge, which have under gone proper aerobic or anaerobic digestion. Sludge from the conventional activated sludge, contacted stabilization, trickling filter, and rotating biological contactor processes usually contain a large amount of volatile solid, which tend to unpleasant odour problem. Therefore this method is generally not suitable for handling this sludge without prior stabilization, and digestion of sludge is essential prior to application of sludge on sludge drying beds.

A typical sludge drying bed consist of 15 to 30 cm of coarse sand layer underlain by approximately 20 to 45 cm of grade gravel ranging in size from 0.6 to 4 cm. Open jointed tubes of 10 to 15 cm diameter spaced at 2.5 to 6 cm are laid in the gravel to provide drainage for liquid passing through the bed. Sludge is applied to the drying bed in layer of 20 to 30 cm, depending upon local climatic conditions the sludge is allowed to dry for two to four weeks. Enclosing drying beds with glass can improve the performance of the dewatering process, particularly in cold or wet climates. For an enclosed bed the area required for a bed may get reduced to two third as compared to area required for open beds.

Tertiary Treatment

This treatment is sometimes called as the final or advanced treatment and consists of removing the organic matter left after secondary treatment, removal of nutrients from sewage, and particularly to kill the pathogenic bacteria. Disinfection is normally carried out by chlorination for safe disposal of treated sewage in water body which is likely to be used at downstream for water supplies. However, for other reuses tertiary treatment is required for further removal of organic matter, suspended solids, nutrients and total dissolved solids as per the needs.

The sewage treatment is generally confined up to secondary treatment only. Various physical chemical and biological processes are available for treatment, depending upon the particular requirements. The choice of treatment methods depends on several factors, including the disposal facilities available. Actually, the distinction between primary, secondary & tertiary treatment is rather arbitrary, since many modern treatment methods incorporate physical, chemical, and biological processes in the same operations.

The secondary treatment can be achieved by aerobic process or anaerobic process. Conventionally the aerobic process i.e. activated sludge process is used for sewage treatment. As a low cost treatment option, oxidation pond can also be used for sewage treatment. With the advent of the energy crises, the use of anaerobic process are being taken into consideration in greater depth as a substitutes for the traditional energy dependent activated process or large area demanding oxidation ponds. The application of anaerobic process for wastewater treatment is attractive only if large volumes of wastewater can be forced through the system in a relatively short period of time. This will give low hydraulic retention time and therefore anaerobic reactor becomes space efficient.

Today majority of wastewater treatment plants use aerobic metabolism for the removal of organic matter. The most well known aerobic processes are the activated sludge process, oxidation ditch, oxidation pond, trickling filter, and aerated lagoons. Stabilization ponds use both the aerobic and anaerobic mechanisms. In the recent years due to increase in power cost and subsequent increase in operation cost of aerobic process, more attention is being paid for the use of anaerobic treatment systems for the treatment of wastewater including sewage. At few places the high rate anaerobic process such as UASB reactor is successfully used for treatment of sewage.

Effluent Quality Requirement

For disposal of treated effluent in the water body or reuse for irrigation the effluent standards are defined by Central Pollution Control Board. For discharge of treated sewage in water body the standard for BOD and SS is 30 mg/L and for application on land for irrigation it is 100 mg/L.

Screens

The primary treatment incorporates unit operations for removal of floating and suspended solids from the wastewater. They are also referred as the physical unit operations. The unit operations used are screening for removing floating papers, rages, cloths, plastics, cans stoppers, labels, etc.; grit chambers or detritus tanks for removing grit and sand; skimming tanks for removing oils and grease; and primary settling tank for removal of residual settleable suspended matter.

Screen is the first unit operation in wastewater treatment plant. This is used to remove larger particles of floating and suspended matter by coarse screening. This is accomplished by a set of inclined parallel bars, fixed at certain distance apart in a channel. The screen can be of circular or rectangular opening. The screen composed of parallel bars or rods is called a rack. The screens are used to protect pumps, valves, pipelines, and other appurtenances from damage or clogging by rags and large objects.

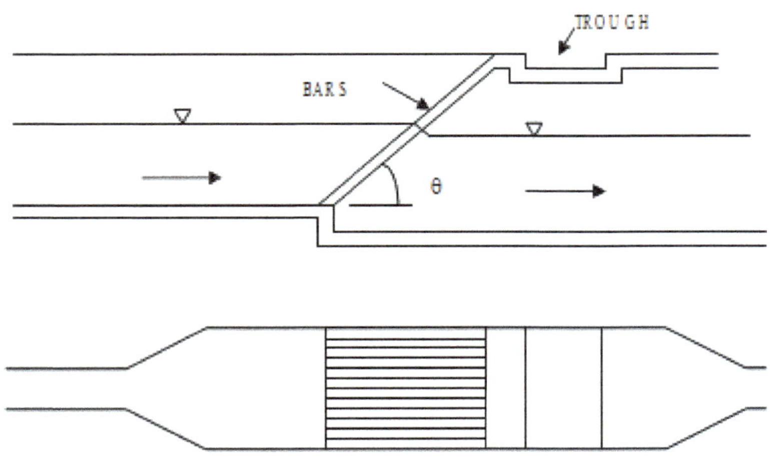

Bar Screen

Industrial wastewater treatment plant may or may not need the screens. However, when packing of the product and cleaning of packing bottles/ containers is carried out, it is necessary to provide screens even for industrial wastewater treatment plant to separate labels, stopper, cardboard, and other packing materials. The cross section of the screen chamber is always greater (about 200 to 300 %) than the incoming sewer. The length of this channel should be sufficiently long to prevent eddies around the screen. The schematic diagram of the screen is shown in the Figure.

Types of Screens

Screens can be broadly classified depending upon the opening size provided as coarse screen (bar screens) and fine screens. Based on the cleaning operation they are classified as manually cleaned screens or mechanically cleaned screens. Due to need of more and more compact treatment facilities many advancement in the screen design are coming up.

Coarse Screen

It is used primarily as protective device and hence used as first treatment unit. Common type of these screens are bar racks (or bar screen), coarse woven-wire screens, and comminutors. Bar screens are used ahead of the pumps and grit removal facility. This screen can be manually cleaned or mechanically cleaned. Manually cleaned screens are used in small treatment plants. Clear spacing between the bars in these screens may be in the range of 15 mm to 40 mm.

Grinder or Comminutor

It is used in conjunction with coarse screens to grind or cut the screenings. They utilize cutting teeth (or shredding device) on a rotating or oscillating drum that passes through stationary combs (or disks). Object of large size are shredded when it will pass through the thin opening of size 0.6 to 1.0 cm. Provision of bye pass to this device should always be made.

Fine Screen

Fine screens are mechanically cleaned screens using perforated plates, woven wire cloths, or very closely spaced bars with clear openings of less than 20 mm, less than 6 mm typical. Commonly these are available in the opening size ranging from 0.035 to 6 mm. Fine screens are used for pretreatment of industrial wastewaters and are not suitable for sewage due to clogging problems, but can be used after coarse screening. Fine screens are also used to remove solids from primary effluent to reduce clogging problem of trickling filters. Various types of microscreens have been developed that are used to upgrade effluent quality from secondary treatment plant. Fine screen can be fixed or static wedge-wire type, drum type, step type and centrifugal screens. Fixed or static screens are permanently set in vertical, inclined, or horizontal position and must be cleaned by rakes, teeth or brushes. Movable screens are cleaned continuously while in operation. Centrifugal screens utilize the rotating screens that separate effluent and solids are concentrated.

Types of Medium and Fine Screens

Inclined (fixed): These are flat, cage, or disk type screens meant for removal of smaller particles.

These are provided with opening of 0.25 to 2.5 mm. They are used for primary treatment of industrial effluents.

Band: It consists of an endless perforated band that passes over upper and lower sprocket. Brushes are installed to remove the material retained over the screen. Water jet can be used to flush the debris. Opening size of 0.8 to 2.5 mm is provided in this screen. They are used for primary treatment of industrial effluents.

Drum Screen or strainer: It consists of rotating cylinder that has screen covering the circumferential area of the drum. The liquid enters the drum axially and moves radially out. The solids deposited are removed by a jet of water from the top and discharged into a trough. The micro-strainers have very fine size screens and are used to polish secondary effluent or remove algae from the effluent of stabilization ponds. Opening size of 1 to 5 mm and 0.25 to 2.5 mm is used for primary treatment and opening size of 6 to 40 μm is used for polishing treatment of secondary effluents.

Screen Chamber

It consists of rectangular channel. Floor of the channel is normally 7 to 15 cm lower than the invert of the incoming sewer. Bed of the channel may be flat or made with desired slope. This channel is design to avoid deposition of grit and other materials in to it. Sufficient straight approach length should be provided to assure uniform distribution of screenings over the entire screen area. At least two bar racks, each designed to carry peak flow, must be provided. Arrangement of stopping the flow and draining the channel should be made for routine maintenance. The entrance structure should have a smooth transition or divergence to avoid excessive head loss and deposition of solids. Effluent structure should be having uniform convergence. The effluent from the individual rack may be combined or kept separate as necessary.

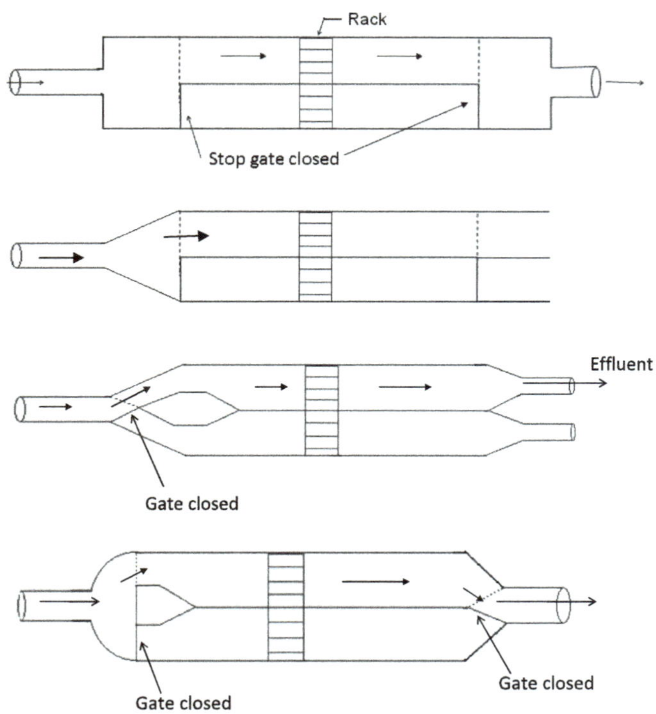

Double chamber bar screen and influent and effluent arrangement

Requirements and Specifications for Design of Bar Screen

1. The velocity of flow ahead of and through a screen varies materially and affects its operation. Lower the velocity through the screen, the greater is the amount of screening that would be removed. However, at lower velocity greater amount of solids would be deposited at the bottom of the screen channel.

2. Approach velocity of wastewater in the screening channel shall not fall below a self cleansing velocity of 0.42 m/sec or rise to a magnitude at which screenings will be dislodged from the bars.

   - The suggested approach velocity is 0.6 to 0.75 m/sec for the grit bearing wastewaters. Accordingly the bed slope of the channel should be adjusted to develop this velocity.

   - The suggested maximum velocity through the screen is 0.3 m/sec at average flow for hand cleaned bar screens and 0.75 m/sec at the normal maximum flow for mechanically cleaned bar screen (Rao and Dutta, 2007). Velocity of 0.6 to 1.2 m/sec through the screen opening for the peak flow gives satisfactory result.

3. Head losses due to installation of screens must be controlled so that back water will not cause the entrant sewer to operate under pressure. Head loss through a bar rack can be calculated by using Kirchmer's equation:

$$h = \beta (W/b)^{4/3} h_v \sin\theta \quad (1)$$

where,

h	= head loss, m	
β	= Bar shape factor	
	= 2.42 for sharp edge rectangular bars	
	= 1.83 for rectangular bars with semicircular upstream	
	= 1.79 for circular bars	
	= 1.67 for rectangular bars with both u/s and d/s faces as semicircular.	
W	= Width of bars facing the flow, m	
b	= Clear spacing between the bars, m	
h_v	= Velocity head of flow approaching the bars, m	
	= $V^2/2g$	
V	= geometric mean of the approach velocity, m/sec	
θ	= Angle of inclination of the bars with horizontal.	

Usually accepted practice is to provide loss of head of 0.15 m but the maximum loss of head with the clogged hand cleaned screen should not exceed 0.3 m. For mechanically cleaned screen, the head loss is specified by the manufacturer, and it can be between 150 to 600 mm.

The head loss through the cleaned or partially clogged flat bar screen can also be calculated using following formula:

$$h = 0.0729 (V^2 - v^2) \quad (2)$$

Where, h = loss of head, m

V = velocity through the screen, m/sec

v = velocity before the screen, m/sec

The head loss through the fine screen can be calculated as:

$$h = (1/(2g.C_d))(Q/A)^2 \quad (3)$$

Where, g = gravity acceleration (m/sec^2); Cd is coefficient of discharge = 0.6 for clean rack; Q is discharge through screen (m^3/sec); and A is effective open submerged area (m^2).

4. The slope of the hand cleaned screen should be in between 30 to 60^0 with horizontal. The mechanically cleaned bar screens are generally erected almost vertical; however the angle with the horizontal can be in the range 45 to 85^0.

5. The submerged area of the surface of the screen, including bars and opening should be about 200% of the cross sectional area of the incoming sewer for separate system, and 300% for the combined system.

6. The clear spacing between the bars may be in the range of 15 mm to 75 mm in case of mechanically cleaned bar screen. However, for the manually cleaned bar screen the clear spacing used is in the range 25 mm to 50 mm. Bar Screens with opening between 75 to 150 mm are used ahead of raw sewage pumping. For industrial wastewater treatment the spacing between the bars could be between 6 mm and 20 mm.

7. The width of bars facing the flow may vary from 5 mm to 15 mm, and the depth may vary from 25 mm to 75 mm. Generally bars with size less than 5 mm x 25 mm are not used. These bars are welded together with plate from downstream side to avoid deformation.

Quantities of Screening

The quantity of screening varies depending on the type of rack or screen used as well as sewer system (combined or separate) and geographic location. Quantity of screening removed by bar screen is 0.0035 to 0.0375 m^3/ 1000 m^3 of wastewater treated (Typical value = 0.015 m^3/1000 m^3 of wastewater) (Metcalf & Eddy, 2003). In combined system, the quantity of screening increases during storm and can be as high as 0.225 m^3/1000 m^3 of wastewater. For industrial wastewaters quantity of the screening depends on the characteristics of the wastewater being treated.

Disposal of Screenings

Screening can be discharged to grinders or disintegrator pumps, where they are ground and returned to the wastewater. Screenings can be disposed off along with municipal solid waste on sanitary landfill. In large sewage treatment plant, screenings can be incinerated. For small wastewater treatment plant, screenings may be disposed off by burial on the plant site.

Example: 1

Design a bar screen chamber for average sewage flow 20 MLD, minimum sewage flow of 12 MLD and maximum flow of 30 MLD.

Solution:

1. Average flow = 20 MLD

   = 0.231 m^3/Sec

   Maximum Flow = 30 MLD

   = 0.347 m^3/Sec

   Minimum flow = 12 MLD

   = 0.139 m^3/Sec

2. Assume manual cleaning and angle of inclination of bars with horizontal as 30^0. Assume size of bars 9 mm x 50 mm, 9 mm facing the flow. A clear spacing of 30 mm between the bars is provided.

3. Assume velocity of flow normal to screen as 0.3 m/sec at average flow.

4. Net submerged area of the screen opening required

$$= \frac{0.231 m^3 / Sec}{0.3 m / sec} = 0.77 m^2$$

Assume velocity of flow normal to the screen as 0.75 m/sec at maximum flow, hence net submerged area of screen opening

$$\frac{0.347 m^3 / Sec}{0.75 m / sec} = 0.46 m^2$$

Provide net submerged area = 0.77 m^2

5. Gross submerged area of the screen

When 'n' numbers of bars are used the ratio of opening to the gross width will be [(n+1)30] / [(n+1)30 + 9 x n] ≈ 0.77 (for 20 to 30 number of bars)

Therefore gross submerged area of the screen 0.77/0.77=1 m^2

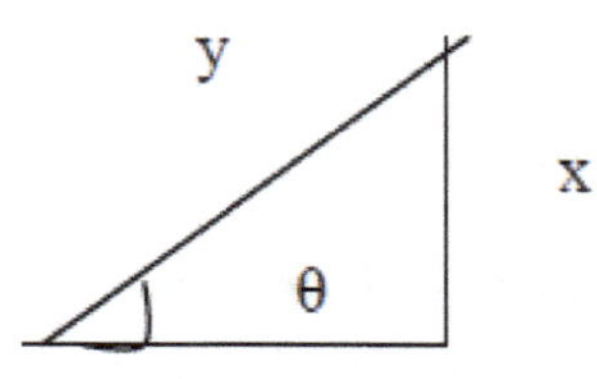

Sin 30 = x/Y

6. The submerged vertical cross sectional area of the screen

$$= 1.0 \times \text{Sin } 30 = 0.5 \text{ m}^2$$

This is equal to c/s area of screen chamber, therefore velocity of flow in screen chamber

$$= 0.231 / 0.5 = 0.462 \text{ m/sec}$$

This velocity is greater than the self cleansing velocity of 0.42 m/sec

7. Provide 30 numbers of bars. The gross width of the screen chamber will be:

$$= 30 \times 0.009 + 31 \times 0.03 = 1.2 \text{ m}$$

Therefore, liquid depth at average flow = 0.5 / 1.2 = 0.416 m

Provide free board of 0.3 m

Hence, total depth of the screen = 0.416 + 0.3 = 0.716 m, say 0.75 m

Thus, the size of the channel = 1.2 m (width) x 0.75 m (depth)

8. Calculation for bed slope:

$$R = A/P = (0.416 \times 1.2) / (2 \times 0.416 + 1.2)$$

$$= 0.246 \text{ m}$$

Now, $V = (1/n) R^{2/3} S^{1/2}$

$$S^{1/2} = V.n / R^{2/3}$$

$$= 0.462 \times 0.013 / (0.246)^{2/3}$$

$$S^{1/2} = 0.0153$$

Therefore bed slope is nearly 1 in 4272 m

9. Head loss through the screen, h, when screen is not clogged.

$$h = \beta (W/b)^{4/3} h_v \text{ Sin } \theta$$

$$= 2.42 (9/30)^{4/3} [(0.462)^2/(2 \times 9.81)] \text{ Sin } 30$$

$$= 2.65 \times 10^{-3} \text{ m} = 0.00265 \text{ m} = 2.65 \text{ mm}$$

For half clogged screen, the head loss can be worked out using opening width as half

Thus, b = 30/2 = 15 mm

And $h = 6.67 \times 10^{-3}$ m = 6.67 mm < 150 mm

However, provide 150 mm drop of after screen.

If this head loss is very excessive, this can be reduced by providing bars with rounded edges at upstream, or by reducing width of bars to 6 to 8 mm, or by slight reduction in velocity. Except for the change in shape of bars in other cases the channel dimensions will change.

For minimum flow and maximum flow, the depth of flow can be worked out using Manning's formula using known discharge, and check for velocity under both these cases, as self cleansing and non-scouring, respectively, and also depth of flow at maximum discharge.

Grit Chamber

Grit chamber is the second unit operation used in primary treatment of wastewater and it is intended to remove suspended inorganic particles such as sandy and gritty matter from the wastewater. This is usually limited to municipal wastewater and generally not required for industrial effluent treatment plant, except some industrial wastewaters which may have grit. The grit chamber is used to remove grit, consisting of sand, gravel, cinder, or other heavy solids materials that have specific gravity much higher than those of the organic solids in wastewater. Grit chambers are provided to protect moving mechanical equipment from abrasion and abnormal wear; avoid deposition in pipelines, channels, and conduits; and to reduce frequency of digester cleaning. Separate removal of suspended inorganic solids in grit chamber and suspended organic solids in primary sedimentation tank is necessary due to different nature and mode of disposal of these solids. Grit can be disposed off after washing, to remove higher size organic matter settled along with grit particles; whereas, the suspended solids settled in primary sedimentation tank, being organic matter, requires further treatment before disposal.

Horizontal Velocity in Flow though Grit Chamber

The settling of grit particles in the chamber is assumed as particles settling as individual entities and referred as Type – I settling. The grit chamber is divided in four compartments as inlet zone, outlet zone, settling zone and sludge zone.

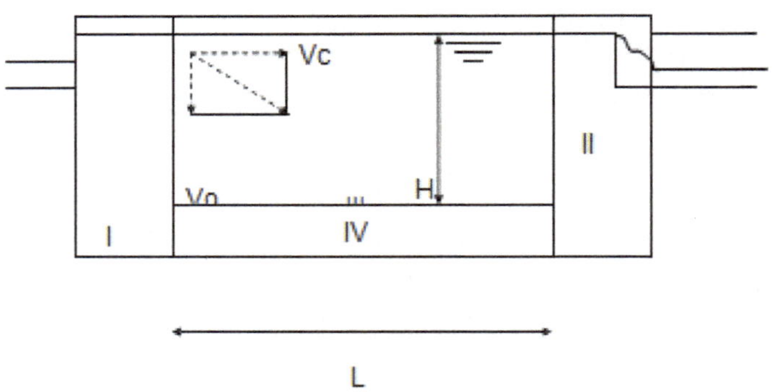

Compartments of grit chamber

Zone – I: Inlet zone: This zone distributes the incoming wastewater uniformly to entire cross section of the grit chamber.

Zone – II: Outlet zone: This zone collects the wastewater after grit removal.

Zone – III: Settling zone: In this zone settling of grit material occurs.

Zone – IV: Sludge zone: This is a zone where settled grit accumulates.

L – Length of the settling zone

H – Depth of the settling zone

v – Horizontal velocity of wastewater

Vo – Settling velocity of the smallest particle intended to be removed in grit chamber.

Now, if Vs is the settling velocity of any particle, then

For Vs ≥ Vo these particles will be totally removed,

For Vs < Vo, these particles will be partially removed,

Where, Vo is settling velocity of the smallest particle intended to be removed. The smallest particle expected to be removed in the grit chamber has size 0.2 mm and sometimes in practice even size of the smallest particle is considered as 0.15 mm. The terminal velocity with which this smallest particle will settle is considered as Vo. This velocity can be expressed as flow or discharge per unit surface area of the tank, and is usually called as 'surface overflow rate' or 'surface settling velocity'. Now for 100 percent removal of the particles with settling velocity Vs ≥ Vo, we have

Detention time = L/v = H/Vo

Or L/H = v/Vo (1)

To prevent scouring of already deposited particles the magnitude of 'v' should not exceed critical horizontal velocity Vc, and the above equation becomes

L / H = Vc / Vo

The critical velocity, Vc, can be given by the following equation (Rao and Dutta, 2007):

$$Vc = \sqrt{\left[\frac{8\beta}{f} g(S-1)D\right]} \quad (2)$$

where, β = constant

= 0.04 for unigranular sand

= 0.06 for non-uniform sticky material

f = Darcy –Weisbach friction factor = 0.03 for gritty matter

g = Gravitational acceleration,

S = Specific gravity of the particle to be removed (2.65 for sand), and

D = Diameter of the particle, m

The grit chambers are designed to remove the smallest particle of size 0.2 mm with specific gravity around 2.65. For these particles, using above expression the critical velocity comes out to be Vc = 0.228 m/sec.

Settling Velocity of the Particles

Settling velocity of any discrete particle depends on its individual characteristics and also on the characteristics of the fluid. Assuming particles to be spherical, the settling velocity of any particle, Vs, can be given by the following formula:

$$Vs = \sqrt{\left[\frac{4}{3}\frac{g}{C_D}(S-1))D\right]}$$

where, C_D = Newton's drag coefficient

$$= \frac{24}{R} + \frac{3}{\sqrt{R}} + 0.34 \quad \text{for } 0.3 < R < 10^4$$

$$= 24 / R, \quad \text{when } R < 0.3$$

R = Reynold's Number = Vs.D/v

v = Kinematic viscosity of the fluid

For the value of R < 0.3, C_D = 24/R and the above equation becomes (Stoke's Law)

$$Vs = \frac{g}{18}\left[\frac{S-1}{v}\right]D^2 \qquad (4)$$

For the value of R > 0.3, the value of Vs should be worked out by trial and error.

Horizontal Flow Rectangular Grit Chamber

A long narrow channel is used in this type of grit chamber. The wastewater moves through this channel in more or less plug flow condition with minimal mixing to support settling of the particles. Higher length to width ratio of the channel is used to minimize mixing. For this purpose a minimum allowance of approximately twice the maximum depth or 20 to 50% of the theoretical length of the channel should be given for inlet and outlet zones. The width of this channel is kept between 1 and 1.5 m and the depth of flow is normally kept shallow. A free board of minimum 0.3 m and grit space of about 0.25 m is provided. For large sewage treatment plant, two or more number of grit chambers are generally provided in parallel. The detention time of 30 to 60 seconds is recommended for the grit chamber.

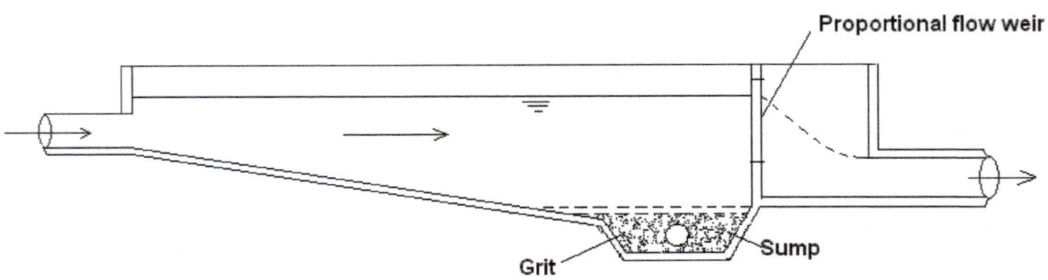

Horizontal flow grit chamber

Control of Velocity Through the Grit Chamber

With variation in sewage flow received at treatment plant, it is important that velocity of the wastewater in the grit chamber should be maintained nearly constant. Otherwise when flow is lower, deposition of not only inorganic solids but also organic solids will occur in grit chamber due to lowering of velocity. With flow higher than average, when the velocity will exceed the critical velocity, scouring of already deposited grit particles will occur leading to failure of performance. Hence for proper functioning, the velocity should not be allowed to change in spite of change in flow in the grit chamber. This can be achieved by provision of proportional weir or Parshall flume at the outlet end of grit chamber. The shape of the opening between the plates of a proportional weir is made in such a way that the discharge is directly proportional to liquid depth in grit chamber. As a result the velocity of water in the chamber will remain constant for all flow conditions.

The discharge through proportional weir can be given by the following equation (Rao and Dutta, 2007):

$$Q = C.b\sqrt{2ag}.\left[H - \frac{a}{3}\right] \qquad (5)$$

where, Q = Discharge, m^3/sec^1

C = constant, 0.61 for symmetrical sharp edged weir

a = 25 to 35 mm as shown in the Figure.

b = base width of the weir

H = Height of water above the crest of weir

The equation of the curve forming the edge of the weir is given by the following formula:

$$x = \frac{b}{2}\left[1 - \frac{2}{\pi}\tan^{-1}\sqrt{\left(\frac{y}{a} - 1\right)}\right] \qquad (6)$$

The sharp edges generated by the curve at the bottom are curtailed on both the side, because such small opening will not contribute for flow due to deposition of solids. These edges are curtailed from the side wall at a distance of minimum 75 mm and height of the vertical edge 'a' is in the range of 25 to 35 mm. To compensate this loss of area the edge of the weir is lowered by a/3 than the theoretical level.

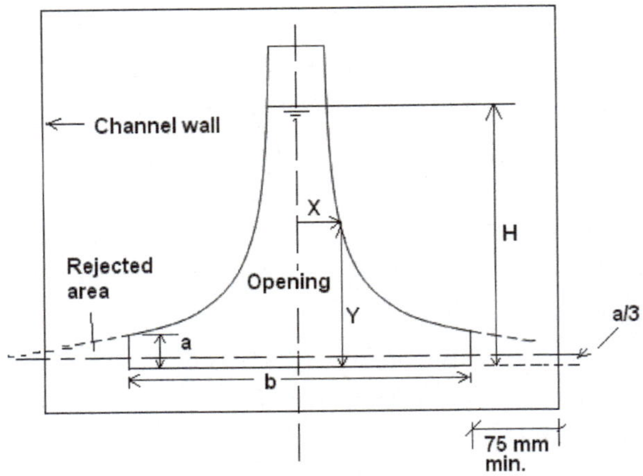

Proportional Weir

Alternatively, Parshall flume can be placed at the end of the grit chamber. The design details for Parshall flume to meet different discharges are provided in the CPHEEO manual (1993). With appropriate arrangement this will also facilitate recording of the discharge received at the sewage treatment plant.

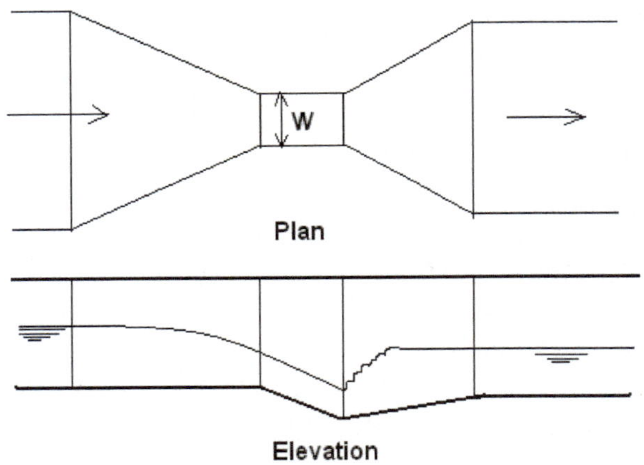

Parshall flume

Disposal of Grit

Considerable quantities of grit will be collected at the sewage treatment plant, about 0.004 to 0.2 m^3/ML. Quantity of grit will be more particularly for combined system. Necessary arrangement should be made at the treatment plant for collection, storage and disposal of this grit matter. The grit collected can be disposed in the following manner:

1. In large treatment plant, grit is incinerated with sludge
2. In the past, grits along with screening was dumped into sea.
3. Generally, grit should be washed before disposal to remove organic matter.
4. Land disposal after washing is most common.

Example:1

Design a grit chamber for population 50000 with water consumption of 135 LPCD.

Solution

Average quantity of sewage, considering sewage generation 80% of water supply, is

$= 135 \times 50000 \times 0.8 \qquad = 5400\ m^3/day = 0.0625\ m^3/sec$

Maximum flow = 2.5 × average flow

$= 0.0625 \times 2.5 = 0.156\ m^3/sec$

Keeping the horizontal velocity as 0.2 m/sec (<0.228 m/sec) and detention time period as one minute.

Length of the grit chamber = velocity × detention time

$= 0.2 \times 60 = 12.0\ m$

Volume of the grit chamber = Discharge x detention time

$= 0.156 \times 60 = 9.36\ m^3$

Cross section area of flow 'A' = Volume / Length = 9.36/12 = 0.777 m^2

Provide width of the chamber = 1.0 m, hence depth = 0.777 m

Provide 25% additional length to accommodate inlet and outlet zones.

Hence, the length of the grit chamber = 12 × 1.25 = 15.0 m

Provide 0.3 m free board and 0.25 m grit accumulation zone depth, hence total depth

$= 0.777 + 0.3 + 0.25 = 1.33\ m$

and width = 1.0 m

Example: 2

Design a horizontal flow grit chamber with rectangular cross section for treating maximum sewage flow of 10 MLD at maximum temperature of 34 ^{o}C during summer and minimum temperature of 15 ^{o}C in winter.

Solution

The settling velocity of the grit particle will be minimum at lower temperature, i.e., 15 ^{o}C. At this temperature kinematic viscosity = $1.14 \times 10^{-2}\ cm^2/sec$

In *first trial* assume Reynolds number 'R' less than or equal to 0.3.

$$Vs = \frac{g}{18}\left[\frac{S-1}{v}\right]D^2$$

$$Vs = \frac{981}{18}\left[\frac{2.65-1}{1.14x10^{-2}}\right]0.02^2$$

$$= 3.15 \text{ cm/sec}$$

Reynolds Number R = v.D/ν = 3.15 x 0.02 / 1.14 x 10^{-2}

= 5.53 > 0.3

Therefore, Vs is not equal to 3.15 cm/sec because the equation for Vs is valid only for R < 0.3. Using Vs = 3.15 cm/sec, calculate R and CD and then again Vs till it converges.

Subsequent Trial

Vs = 2.4 cm/sec

R = 2.4 x 0.02 /(1.14 x 10^{-2}) = 4.21

$$C_D = \frac{24}{4.21} + \frac{3}{\sqrt{4.21}} + 0.34$$
$$= 7.50$$

From equation

$$V_s = \sqrt{\left[\frac{4}{3}\frac{981}{7.50}(2.65-1)0.02\right]}$$

Vs = 2.4 cm/sec Hence, O.K. (2074 m/d)

Now for β = 0.06, f = 0.03, and D = 0.02 cm

$$Vc = \sqrt{\left[\frac{8\beta}{f}g(S-1)D\right]}$$

$$Vc = \sqrt{\left[\frac{8x0.06}{0.03}981(2.65-1)0.02\right]}$$

$$= 22.76 \text{ cm/sec}$$

Now Q = 10 MLD = 0.116 m^3/sec

Therefore, C/S Area A = Q/V = 0.116/0.227 = 0.51 m^2

If width of 1 m is provided, the depth required = 0.51 m

Provide total depth = 0.51 + 0.3 (free board) + 0.25 (space for grit accumulation)

= 1.06 Say 1.1 m

Now Vo/Vc = H/L = 2.4/22.7

Therefore theoretical length L = 22.7 x 0.51 / 2.4 = 4.824 m

Provide 2 m extra length for inlet and outlet

Therefore total length = 2 + 4.824 = 6.824 m say 6.9 m

Total working volume = 0.51 x 6.9 x 1 = 3.52 m^3

Hence, Overall detention time = 3.52 / 0. 116 = 30.34 sec (within 30 to 60 seconds)

Square Grit Chamber

The horizontal flow rectangular grit chamber faces the problem of sedimentation of organic matter along with grit particles, requiring external washing of the grit before disposal. This problem can be minimized by providing square shape of the grit chamber rather than long rectangular channel. Also, this shape will facilitate compact design of sewage treatment plant. Hence these days' square grit chambers are used. In square grit chamber, the flow distribution may not be uniform due to non-ideal plug flow conditions, and hence continuous removal of grit is generally considered essential. These are designed based on overflow rates that are dependent on the particle size and temperature of wastewater. Minimum two number of grit chambers should be used to facilitate maintenance of the raking mechanism, whenever required. The grit deposited at the bottom is raked by rotating mechanism to a sump at the side of the tank, from which it is moved up by an inclined reciprocating rake or screw pump mechanism. While passing up the incline conveyer, organic solids are separated from grit and flow back into the basin. Thus, cleaned washed grit is obtained, compared to the grit obtained from separate grit washers.

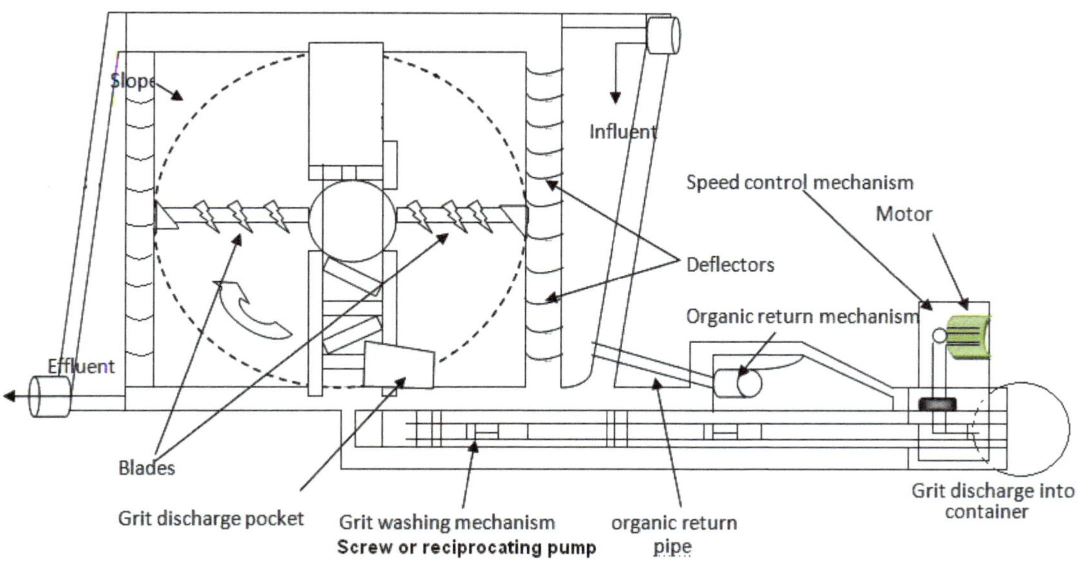

Square grit chamber

Aerated Grit Chamber

Excessive wear of grit handling equipment and necessity of separate grit washer can be eliminated by using aerated grit chamber. It is designed for typical detention time of 3 minutes at maximum flow. Grit hopper of about 0.9 m deep with steeply sloping sides is located along one side of tank under air diffusers. The diffusers are located at about 0.45 to 0.6 m from the bottom. The size of particles removed will depend upon velocity of roll or agitation. The air flow rate can be easily adjusted to control efficiency and 100% removal of grit can be achieved. Wastewater moves in the tank in helical path and makes two or three passes across the bottom of the tank at maximum flow (and more at less flow). Wastewater is introduced in the direction of roll in the grit chamber. The expansion in volume due to introduction of air must be considered in design. The aerated grit chambers are equipped with grit removal grab buckets, traveling on monorails over the grit collection and storage trough. Chain and bucket conveyers can also be used. Two grit chambers in parallel are used to facilitate maintenance. Typical design details for aerated grit chamber are provided below (Metcalf and Eddy, 2003):

Depth : 2 to 5 m

Length : 7.5 to 20 m

Width : 2.5 to 7.0 m

Width to depth ratio: 1:1 to 5:1

Detention time at peak flow: 2 to 5 min (3 minutes typical)

Air supply m^3/min.m of length : 0.15 to 0.45 (0.3 typical)

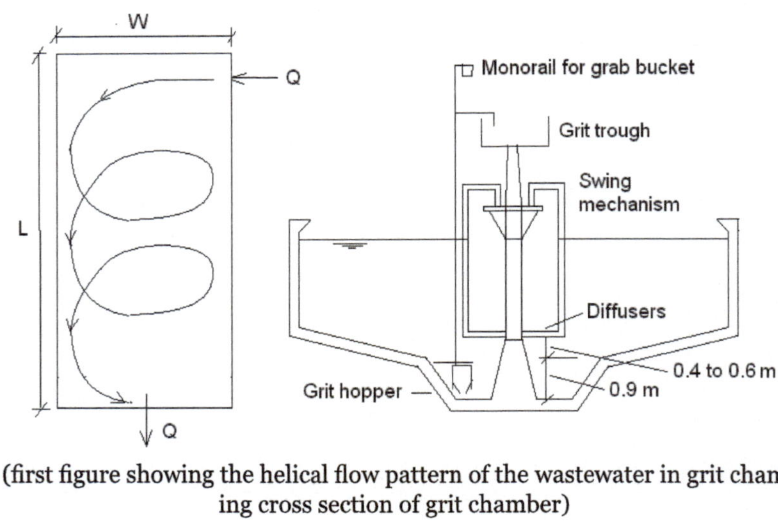

Aerated grit chamber (first figure showing the helical flow pattern of the wastewater in grit chamber and second showing cross section of grit chamber)

Example: 3

Design aerated grit chamber for treatment of sewage with average flow of 60 MLD. Consider the peak factor of 2.

Solution:

1. Average flow = 60 MLD = 0.694 m^3/sec, and Peak flow = 0.694 x 2.0 = 1.389 m^3/sec
2. Volume of grit chamber

Provide two chambers to facilitate periodic cleaning and maintenance Provide detention time = 3.0 min

Volume of each tank = 1.389 x 3 x 60 /2 = 125.01 m^3

3. Dimensions of aeration basin:

   Provide depth to width ratio of 1: 1.2

   Provide depth = 3.0 m, hence width = 1.2 x 3.0 = 3.6 m

   Length = 125.01 / (3 x 3.6) = 11.575 m

   Increase length by 20% to account for inlet and outlet conditions.

   Total length = 11.575 x 1.2 = 13.89 m.

4. Determine the air-supply requirement

   Consider 0.3 m^3/min.m of length air supply

   Air Requirement = 13.89 x 0.3 = 4.17 m^3/min

   Provide air swing arrangement at 0.5 m from floor

5. Quantity of grit :

   Consider grit collection 0.015 $m^3/10^3$ m^3

   Volume of grit = 1.389 x 60 x 60 x 24 x 0.015 x 10^{-3}

   = 1.8 m^3/d

6. Check for surface overflow rate (SOR)

   The settling velocity of the smallest particle = 2.4 cm/sec, the actual SOR in the grit chamber = 1.389 / (2 x 3.6 x 11.575) = 0.0167 m/s = 1.67 cm/sec, which is less than the settling velocity of the smallest particle hence design is safe.

Vortex-Type Grit Chamber

These types of grit chambers are used in small plants and these require lesser area as compared to earlier types. In this type, grit is removed with vortex flow pattern. The wastewater enters tangentially and exit in the perpendicular direction of motion either from top or from side. Due to inertia the grit particles will remain in the chamber and liquid free from grit will only escape. The rotating turbine maintains constant velocity and helps in separating organic matter and grit. The centrifugal force on the grit particle can also be maintained without turbine by properly introducing wastewater in the tangential direction in the chamber. Toroidal flow path is followed by the grit particles due to action of propeller (Metcalf and Eddy, 2003). Grit particle settles by the action of gravity into hopper from where it is removed by a grit pump or air lift pump. Washed grit, free from the organic matter, can be obtained from this device.

References

- "Lowering Cost and Waste in Flue Gas Desulfurization Wastewater Treatment". Power Mag. Electric Power. Retrieved 6 April 2017
- Margot, J.; et al. (2013). "Bacterial versus fungal laccase: potential for micropollutant degradation". AMB Express. 3: 63. doi:10.1186/2191-0855-3-63
- Kashiwaya, Mamoru; Yoshimoto, Kameo (May 1980). "Tannery Wastewater Treatment by the Oxygen Activated Sludge Process". Journal (Water Pollution Control Federation). Alexandria, VA: Water Environment Federation. 52 (5): 999–1007. JSTOR 25040825
- Sharma, Sanjay Kumar; Sanghi, Rashmi (2012). Advances in Water Treatment and Pollution Prevention. Springer. ISBN 9789400742048. Retrieved 2013-02-07
- "Waste Pesticide Management" (PDF). Oregon.Gov. State of Oregon Department of Environmental Quality Land Quality Division Hazardous Waste Program. Retrieved 5 October 2016
- Haughey, A. (1968). "The Planktonic Algae of Auckland Sewage Treatment Ponds". New Zealand Journal of Marine and Freshwater Research. 2 (4): 721–766. doi:10.1080/00288330.1968.9515271
- Natural Environmental Research Council – River sewage pollution found to be disrupting fish hormones. Planetearth.nerc.ac.uk. Retrieved on 2012-12-19
- Ashton, John; Ubido, Janet (1991). "The Healthy City and the Ecological Idea" (PDF). Journal of the Society for the Social History of Medicine. 4 (1): 173–181. doi:10.1093/shm/4.1.173. Retrieved 8 July 2013
- Tilley, David F. (2011). Aerobic Wastewater Treatment Processes: History and Development. IWA Publishing. ISBN 9781843395423. Retrieved 2013-02-07
- Logan, B.; Regan, J. (2006). "Microbial Fuel Cells—Challenges and Applications". Environmental Science & Technology. 40 (17): 5172–5180. doi:10.1021/es0627592

Permissions

All chapters in this book are published with permission under the Creative Commons Attribution Share Alike License or equivalent. Every chapter published in this book has been scrutinized by our experts. Their significance has been extensively debated. The topics covered herein carry significant information for a comprehensive understanding. They may even be implemented as practical applications or may be referred to as a beginning point for further studies.

We would like to thank the editorial team for lending their expertise to make the book truly unique. They have played a crucial role in the development of this book. Without their invaluable contributions this book wouldn't have been possible. They have made vital efforts to compile up to date information on the varied aspects of this subject to make this book a valuable addition to the collection of many professionals and students.

This book was conceptualized with the vision of imparting up-to-date and integrated information in this field. To ensure the same, a matchless editorial board was set up. Every individual on the board went through rigorous rounds of assessment to prove their worth. After which they invested a large part of their time researching and compiling the most relevant data for our readers.

The editorial board has been involved in producing this book since its inception. They have spent rigorous hours researching and exploring the diverse topics which have resulted in the successful publishing of this book. They have passed on their knowledge of decades through this book. To expedite this challenging task, the publisher supported the team at every step. A small team of assistant editors was also appointed to further simplify the editing procedure and attain best results for the readers.

Apart from the editorial board, the designing team has also invested a significant amount of their time in understanding the subject and creating the most relevant covers. They scrutinized every image to scout for the most suitable representation of the subject and create an appropriate cover for the book.

The publishing team has been an ardent support to the editorial, designing and production team. Their endless efforts to recruit the best for this project, has resulted in the accomplishment of this book. They are a veteran in the field of academics and their pool of knowledge is as vast as their experience in printing. Their expertise and guidance has proved useful at every step. Their uncompromising quality standards have made this book an exceptional effort. Their encouragement from time to time has been an inspiration for everyone.

The publisher and the editorial board hope that this book will prove to be a valuable piece of knowledge for students, practitioners and scholars across the globe.

Index